NON FACET VER.

VERIS tempore florent
is dulcedine mundus.

Picturing PLANTS

Tab. 4.

Classis III. TRIANDRIA. A. Monogynia. a. Floribus spathaceis triloculares.

83 Crocus. Saffran.

415. VI. 3. 5. DENUDATE et varia SAFFRAN XR.
Colchicum. Zeitlosen. Licht-Blume.
Bulbocodium Gelbe Licht-Blume.
366. VI.

Xyris. gras-Lilie. 359.

Aphyllanthes Blaue Nb.

Eriocaulon. Randia Knopf Gras. 95. 1. III. 1. 2.

Ixie Sp.

Antheria Sp.

Ixia. Ixie. Chineser Lilie. 54.

2. ENSATE SCHWERTEL KRÆUTER.
Gladiolus. Feder-Lilie. 55.

Antholyza. Æthiopische Lilie. 56.

Comelina. Comeline. 58.

Bermudiana.
Bermudische Lilie.
908. Sisyrinchium.
XX 3.

v. T. 64.

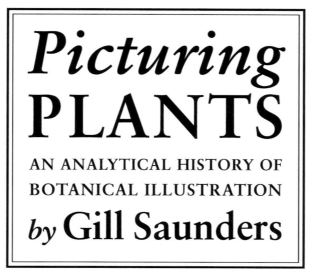

Picturing PLANTS

AN ANALYTICAL HISTORY OF BOTANICAL ILLUSTRATION

by Gill Saunders

UNIVERSITY OF CALIFORNIA PRESS

Berkeley   Los Angeles

in association with

THE VICTORIA AND ALBERT MUSEUM

London

Frontispiece Detail from Johann Gesner, Class III. Triandria (pl. 66)

Front cover Paeonies. Plate from the Nassau Florilegium *c.*1650-70
by Johann Jakob Walther (pl. 25)

Back cover Violet. Plate from *Herbier de la France,* Paris, 1780-95
by Pierre Bulliard (pl. 23)

University of California Press
Berkeley and Los Angeles, California

Text and illustrations © the Trustees of the Victoria and Albert Museum

First published in 1995 in the United Kingdom by Zwemmer
and in the United States by University of California Press

ISBN 0-520-20306-2

Cataloging-in-Publication Data on file with the Library of Congress

Designed and typeset by Christopher Matthews and Ian Muggeridge
Printed and bound in Singapore

9   8   7   6   5   4   3   2   1

Contents

Preface

Since its foundation in 1856 the Victoria and Albert Museum has collected examples of botanical illustration in all graphic media. These range in date from the fifteeenth century to the present day. Whilst the Natural History Museum collects botanical prints as an extension of the herbarium specimen collections, the V&A's purposes in acquiring such material are more complex and varied. The early herbals, for instance, exemplify the woodcut process as applied to the art of the book, whilst the printed and painted florilegia have been used as pattern books by artists and designers such as William Morris. The archives of horticultural journals chart fashions in garden plants, but also demonstrate the progress of colour-printing. Even that great treasure of early botanical illustration, the album of plant studies by Jacques Le Moyne de Morgues, was purchased in 1856 not for its artistic or botanical merit but for the fine late-sixteenth-century binding. The significance of the drawings themselves in the history of botany was not recognized until 1922.

This book looks at botanical illustration from the perspective of an art museum, which inevitably privileges artist over subject, and aesthetics over scientific value. It is not intended as a comprehensive general history of the subject; that has been well done elsewhere (see bibliography). Rather it sets out to explore the ways in which art and science are interdependent in botanical illustrations, and to analyze the graphic languages that have been applied to the representation of botanical subjects. It will describe the different uses to which botanical illustrations have been put, and the ways in which the style and content of the image are the products of graphic technologies, of purpose and context. In the process it will discuss and illustrate herbals (works describing utilitarian, mostly medicinal, plants), florilegia (decorative flower books, often associated with a particular garden), floral pattern books, treatises on taxonomy, botanical monographs (works devoted to a single family of plants), horticultural literature, floras and field guides.

Introduction

1 Ray, J. *Correspondence*, 1848, p.155.

In the words of the seventeenth-century English botanist John Ray, many people 'looked upon a history of plants without figures as a book of geography without maps'.[1] Indeed we might see botanical illustrations as 'mapping' our knowledge of plants. Like cartography, they progress from fanciful and inaccurate images derived from debased records of received knowledge, which draw on myth, rumour and hearsay, to increasingly analytical images that are the product of direct experience and observation. And the ultimate applications of this process, floral diagrams and dissections, and the products of microscopy and x-ray, can be considered as literally maps of internal structure.

Perhaps more than any other scientific discipline botany has been dependent on illustrations for its development. The illustration stands as a substitute for the thing itself, which is ephemeral, fragile, and often unable to survive removal from its original environment. Unlike insects, stuffed birds or animal skeletons, plants preserved as dried herbarium specimens lose much that is essential to their character, such as volume, plasticity and colour, and are thus of limited value as a currency for the exchange of botanical knowledge. Redouté explained his special concern for the colouring of the plates in his great work *Les Liliacées* (1802-16):

> ...it is not only for the pleasure of the eyes that I have undertaken the...work; naturalists have long regretted their inability to conserve the Liliaceae [a genus of plants which included such fragile and fleshy kinds as fritillaries, crocuses and irises] in their herbaria; the accuracy of the descriptions and the fidelity of the engraving will save them the trouble of trying.

Pictures of plants have been the medium for identification, analysis and classification; they have served the physician and pharmacist, the botanical scientist, the taxonomist, the plant collector, the gardener, the designer of applied arts, and the amateur enthusiast of 'natural history'. The value of illustrations is attested by several writers including Robert Wright, the author of important Indian floras; writing to the *Madras Journal of Science* (15 October 1837) he observed that 'the insufficiency of language alone, to convey just ideas of the forms of natural objects, has led naturalists, ever since the invention of engraving, to have recourse to pictorial delineation to assist the mind through the medium of the senses.'

In the work of those illustrators most admired by botanists and by writers on illustration, almost the entire burden of the publication may be carried by the pictures; this finds its apotheosis in the hand-coloured engravings in Franz Bauer's *Delineations of Exotick Plants* (1796-1803). This compilation of plates illustrating South African ericas collected for Kew by Francis Masson has no text beyond a preface by Sir Joseph Banks who explains that 'it would have been a useless task to have compiled, and superfluous expense to have printed, any kind of explanation concerning [the plates]; each figure is intended to answer itself every question a Botanist can wish to ask, respecting the structure of the plant it represents…'

But this was exceptional. Generally speaking, only works with a primarily decorative intent – florilegia and pattern books – are able to dispense with text, whereas illustrations with a botanical purpose are both dependent on the text for interpretation and expansion, and serve in their turn to elucidate the written data. Description alone is rarely sufficient for proper identification, and certain essential features can only be described unambiguously by means of a picture. As Eugenio Battisti observed[2], in the sixteenth and seventeenth centuries many plant descriptions follow from the illustration rather than preceding it; and later Linnaeus found published illustrations invaluable as the 'types' of new species. Significantly, when a universal technical language of description and classification was established in the eighteenth century, many works of botanical theory dispensed with illustrations altogether, though they remained essential for books of a practical, descriptive, decorative or reference nature. Again in the twentieth century, verbal description, rather than illustration, has been seen as the principle of objective science, and as a consequence a major field guide – the 579-page *Excursion Flora of the British Isles* (Clapham, Tutin and Warburg, Cambridge, 1959) – could appear without a single illustration.

2 *L'Antrinasciamento*, Milan, 1962, p.62.

The value of illustrations to botanical or horticultural writing might seem self-evident but there were dissenting voices criticizing what they perceived to be inadequate representations of plants: John Rea, in the preface to his gardening manual/florilegium *Flora: seu, De Florum Cultura* (1676) asserted:

> As for the Cutting of Figures for every Plant, especially in Wood, as Mr. Parkinson hath done [in his *Paradisus in Sole*, 1629], I hold to be altogether needless, such Artless things being good for nothing unless to raise the Price of the Book; serving neither for Ornament or Information, but rather to puzzle or affright the Spectators into an aversion…

His objections, though expressed in exaggerated terms, are to some degree justified; the woodcut, still at this date a common method of reproduction, was generally a clumsy medium for the kind of fragile details that characterize flowering plants. The reputations of the Fuchs and Brunfels herbals resides very largely in the exceptional quality of their illustrations, their unusual success in transcending some of the limitations of the woodcut, and their shift towards naturalism.

The picturing of plants has fluctuated between the general and the specific, the earliest printed herbals representing plants simplified and stylized to the point of distortion. The illustrations in the *Latin Herbarius* (1484) or *Ortus Sanitatis* (*c*.1500) are crude diagrams, ciphers – they represent a folkloric 'idea' of the plant and have the direct simplicity of hieroglyphs or ideograms. A pictorial identity is of so little account that the same woodcut image may be used to 'illustrate' several different plants. Such images belong to a period before the systematic study of plants, and before there was a perceived value to direct observation. They were also the product of a culture in which images were secondary to words as a source of knowledge. The exactly-reproducible image was a relatively recent invention[3], and although it was to transform ways of seeing, it was not yet a primary carrier of knowledge about the world. Nor was the audience for these publications experienced in the 'reading' and interpretation of visual imagery. Text was privileged, and images were marginalized, subordinate. They acted merely as codifications of existing knowledge about plants, which was partial, often inaccurate. They were the latest stage in a series of copies of Classical manuscript illustrations which were accepted without question as accessories to authoritative texts. Each copy is a further distortion, debased by translation into the crude linear medium of the woodcut. By this process the image is reduced to the point where it is virtually useless for purposes of precise identification, and even the most distinctive of plants is represented by a set of basic signifiers – the rose as its own heraldic likeness, the pine with its needle-like leaves rendered as a tight zig-zag band, the oak with its outsize acorns dwarfing a tree ludicrously out of scale, the paeony with its lumpy tuberous root and globular flowers.

In contrast to these, Weiditz's drawings for Brunfel's *Herbarum Vivae Eicones* (1530-36) are precise portraits of individual plants, wilting and damaged, as are the naturalistic examples amongst Le Moyne's modest but exquisite studies of common native plants. These two sixteenth-century artists emphasized the particular, but the major shift towards generalized

3 William Ivins, in *Prints and Visual Communication*, Cambridge, Mass. and London, 1953, suggests a date of around 1400, although there is no evidence for the regular use of prints in books until the 1460s.

Leonurus Canadensis Origani folio. Tourn.

1 Georg Dionysus Ehret (1708-70)

Oswego tea-plant (*Monarda didyma*)

Watercolour, bodycolour and gold leaf

Until his visit to Paris in the winter of 1734-5, Ehret had always painted on paper. However the studies he sent to Trew during his stay at the Jardin Royal des Plantes were, as Trew noted in his *memoir*, painted on 'pergament' (vellum). It seems likely that this change in practice was a direct consequence of seeing the *velins du roi*, the great collection of botanical portraits on vellum painted for the king. The series began with Nicolas Robert (1614-85) and continued with contributions from the succeeding official painters, including Aubriet, Madeleine Basseporte, and Gerard von Spaendonck. Basseporte was in the post at the time of Ehret's visit and probably instructed him in these new skills.

This plate is exceptional in Ehret's oeuvre because it has a gold leaf border, a standard feature of the *velins*. It seems to have been an isolated attempt to imitate their style of presentation.

2 Pierre Joseph Redouté (1759-1840)

Canterbury Bells (*Campanula medium*)

Watercolour on vellum

This is an informal study made as a gift for James Lee (1715-95), the botanist, whilst Redouté was staying at his home in Hammersmith in 1787 (the picture is signed and dated). Though not intended for publication it is nevertheless a *tour de force* of technical and aesthetic skill. At the age of only 28, Redouté had perfected a style which exploited the transparency of watercolour washes applied to the luminous smoothness of vellum. The sheen of the petals is rendered so convincingly that we can 'read' the texture as well as if the plant itself were laid fresh on the page. The leaves are painted in a slightly artificial manner, in the palest shades of bluish-green.

In the tradition of northern European flower painting he includes two insects and by doing so makes a grand claim for the life-likeness of his picture. We are reminded of Pliny's story of the Greek painter Zeuxis who painted grapes so realistically that they fooled birds who flew down to peck at them.

representation came at the same period, with Fuchs' *De Historia Stirpium* (1542). In this work, the woodcut illustrations are careful schematic depictions of the generic forms, but based on observation of the thing itself and not on a second-hand motif. There is no longer a place for individuality or accident. The illustrations of Ehret, Redouté and many others give us idealized specimens, even though they often cite a specific plant as their subject (as Ehret does with the Magnolia grandiflora and the Turk's cap lily in the V&A). Often this image of the ideal plant is invented to the degree that it combines in the one image different stages of flowering and fruiting that would not occur concurrently but are essential parts of its identity (as in the work of Fuchs' illustrator). In more recent times, the techniques of nature printing and photography have returned to the depiction of an individual specimen complete with the peculiarities of colour and form, and the marks of disease and damage specific to that one unique example, and of course the seasonal particularity of its moment of representation. Thus Roger Phillips, in his field guides, must with certain plants show the same species at different times of year, in order to represent the full range of its distinguishing features. Photography has never superseded drawing in the field of botanical illustration because an artist can with greater economy and precision combine these features, and highlight or focus on those of particular value for purposes of identification.

An illustration is necessary to advance knowledge, but at the same time an artist will draw what he knows. As astronomer Charles Piazzi Smyth demonstrated with his photographs of the dragon tree of Tenerife, its characteristic structure had never been accurately represented by European artists since an anonymous watercolour was reproduced as a woodcut in the herbals of Charles de L'Ecluse and Matthias de L'Obel (both published in 1576). This was a reasonably accurate if over-symmetrical account of the tree's form, but representations in later centuries introduced errors and distortions in ostensibly more sophisticated images. This was because the artists came to the subject with the European concept of 'tree' and could only show it as a variant form of the familiar model. In other words, what they saw was functionally limited by their experience. How and what we see depends crucially on what we know.

A comparison between Piazzi Smyth's photographic records and earlier drawn records of the dragon tree shows that even first-hand and supposedly uninflected representations of an unfamiliar plant are subject to cultural influences. The unknown is configured in terms of the known. We see this

3 Paul de Reneaulme

Carnation (*Dianthus caryophyllus*)

Plate from *Specimen Historiae Plantarum*, Paris 1611

Etching

Reneaulme (1560-1624) was a physician. His book, the product of his botanizing in the Alps, Italy and the environs of Paris, is modest, containing only 25 plates etched by the author. However it was of sufficient botanical significance to be later cited by Linnaeus. The illustrations are extraordinary for the date: finely detailed naturalistic studies that recall Jacques Le Moyne de Morgues. Each plate shows the whole plant but with additional studies of variant forms. The book is an excellent demonstration of how apt etching, with its fine sensitive line, is to the demands of botanical illustration.

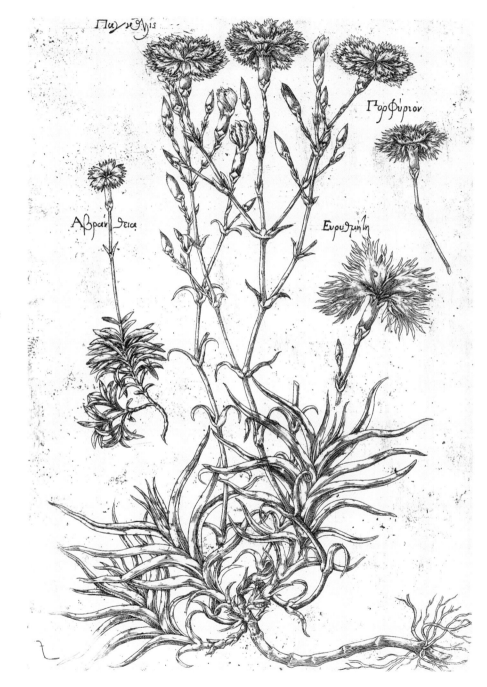

again in the eighteenth- and nineteenth-century drawings made by Chinese and Indian artists attempting to follow European pictorial conventions: the so-called 'stiffness' and awkwardnesses in these drawings are the consequence of the artists' inability to wholly assimilate a foreign style at odds with their own practice and their own innate ways of seeing. Throughout the history of botanical illustration we find the character of a representation shaped by the re-casting of the unfamiliar in terms of the familiar. The objective reality is always subverted by local and national differences in perception, and in the graphic conventions and techniques available.

We find too that the representation of a plant depends on a series of choices about what is relevant and valuable and what is not. As botanical science has developed it has dictated to the artist as to what an illustration should include, and how it should represent what is included. Roots and underground parts of plants were often of medical value to the herbalist and the apothecary, and knowing this, the artist illustrating a herbal invariably gives them as part of his account of the plant. The flower, whilst always part of the plant's identity, does not enjoy a singular focus until the emergence of classification systems which use the structure of the flower as their theoretical basis.

The detail and accuracy of an illustration may be circumscribed by technical factors – the inability of the woodcut to provide fine line and subtle detail; the limited palette of pigments available to the colourist in the sixteenth and seventeenth centuries; by the facility and expertise of the artist and the block-cutter or engraver – but it is also limited by the syntactical schemes available to the artist. Thus just as the same perception will have a distinctive expression in the visual description of individual artists, so even greater differences will exist between visual representations of the same subject produced at different times or in different cultures. Europeans tend to see Western botanical illustration as 'true to nature', and to see a more emphatic two-dimensional decorative aspect to illustrations made in China and Japan. Goethe made this distinction in praising Ferdinand Bauer for his skilful grading of tones in his illustrations for Lambert's *Description of the Genus Pinus* (1803-42) because they were shown in their true spatial relationship and not simply reduced to the flat pattern Goethe considered to be characteristically 'Chinese'.[4] He says that in these plates 'Nature is visible, Art is concealed'. Such judgements are not objective, but depend upon our familiarity with a certain set of codes and conventions governing pictorial representations.

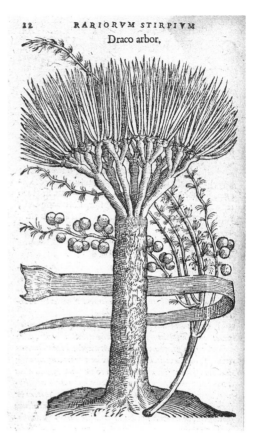

4 Dragon tree (*Dracaena draco*)

From Matthias de L'Obel, *Plantarum seu Stirpium Historia*, 1576

Woodcut

This was one of the blocks collected by Christophe Plantin, the Antwerp printer and publisher, and used as common stock. It is a more accurate account of the tree than any subsequently published in Europe, until photography established an objective truth unmediated by cultural conventions.

4 Quoted in Blunt, W. and Stearn, W. T. *The Art of Botanical Illustration* revised ed, 1994, p.231-3.

5 Charles Piazzi Smyth (1819-1900)

Young dragon trees (*Dracaena draco*)
near Orotava

Plate from *Teneriffe*, London, 1858

Photo-stereograph

In a letter to John Hutton Balfour, 24
February 1859 (in the Botanic Gardens,
Edinburgh) Piazzi Smyth says 'The old
drawings of this tree, you know, are so
useless for truth and science, and now a new
one has just appeared in Germany by Doctor
Hermann Schatt, correcting some of the old
errors but introducing some new ones which
are frightful. But the dragon will never be
"slayed" until photography is employed in
taking his portrait and printing it in books.'

The history of the representation of the
dragon tree offers a perfect example of
photography's role in re-viewing subjects,
and correcting long-held misconceptions.
Though photography has not supplanted the
botanical drawing, it has influenced ways of
seeing and contributed to a revision of
scientific imagery.

5 Quoted in William M. Smallwood and Mabel S.C.
Smallwood, *Natural History and the American Mind*, New
York, 1941, p.26.

6 See Blunt, W. *The Art of Botanical Illustration*, 1950,
p.203 for Thornton, and p.237 for North.

Just as there are two conventions of representation - the particular and the
generic – so there are two basic styles in botanical illustration (common to
all natural history drawings), the illusionistic pictorial representation, and
the outline schematic representation; the former is generally coloured, the
latter plain, though this is not an infallible rule. Mark Catesby, the illustrator of *The Natural History of Carolina, Florida and the Bahama Islands*
(1731-43), expressed the general view when he wrote, 'Things done in a
Flat, tho' exact manner, may serve the Purpose of Natural History, better in
some Measure, than in a more bold and Painter-like Way'.[5] A coloured
image conveys different information from that given by a black and white
outline; it does not necessarily follow that the former is better than the latter, because value is bound up with purpose and whilst colour may be essential for the purposes of the gardener or the amateur botanist, an uncoloured
line-drawing, with its clarity and detail, will often be of more use in the
analysis of plant morphology.

One convention that is more or less standard in botanical illustration is the
de-contextualisation or isolation of the specimen on a blank ground.
Deviations from this mode of presentation are often deemed unscientific –
Thornton's *Temple of Flora* (1799-1807) for example, or Marianne
North's oil studies of exotic plants in their native habitats[6] (Marianne
North Gallery, Kew). Indeed only field guides, as their name indicates,
attempt to relate the plants depicted to a landscape, an appropriate and

accurately-depicted physical context. Gardening books, too, more especially since photography became their dominant illustrative medium, are interested not only in the depiction of individual plants, but in the relation of those plants to their setting, and to each other.

In order to fully understand and interpret botanical imagery it is necessary to recognize and understand the graphic languages employed; these are themselves dictated by the purpose of the image, and by the technology employed in its production or reproduction. Even today the illustrator using traditional techniques remains indispensable in the production of a scientifically complete portrait of a plant, working often piecemeal from fragmentary evidence, or from a preserved herbarium specimen perhaps supplemented by sketches and colour notes, and using a combination of experience and imagination to construct the whole from the part. The demonstrative picture is still vital to botany and horticulture, whatever form it takes, be it a drawing with colour notes, a fully-realized plate for reproduction in a botanical journal, a picture to identify the contents of a seed packet, or a photomicrograph of a plant's molecular structure.

Herbals

The earliest English printed book about plants – an anonymous unillustrated volume printed by Richard Banckes of London in 1525 – defines the herbal as a book which 'sheweth and treateth of ye vertues and proprytes [properties] of herbes'. As this suggests, the herbal had its foundations in medicine and the utilitarian interest in plants. Though the Greek writer Theophrastus (372-287 BC) had stated that botany should not be preoccupied with medicinal or practical virtues but 'must consider the distinctive characters and general nature of plants from the standpoint of their morphology, their behaviour in the face of external conditions, their mode of generation and their whole manner of living', botany was not established as a science distinct from medicine until the later sixteenth century in Europe. Hence plant lore and the study of plants were concerned with their medical properties, and the first gardens devoted to the collection and study of plants were physic gardens, administered not by botanists but by physicians or apothecaries. It was these men who wrote or compiled the first herbals. And it is in the herbals, in the form of manuscripts and later, printed books, that the first botanical illustrations appear – pictures of plants with the practical purpose of offering aids to identification.

Pliny the Elder in his *Natural History* (written in the first century AD) tells us that the Greek herbals of the first century BC had coloured illustrations of the plants (though he had reservations about their reliability):

> For they [ie. Krateuas, Dionysius, and Metrodorus] painted likenesses of the plants and then wrote under them their properties. But not only is a picture misleading when the colours are so many, particularly as the aim is to copy Nature, but besides this much imperfection arises from the manifold hazards in the accuracy of the copyists. In addition, it is not enough for each plant to be painted at one period only of its life, since it alters its appearance with the fourfold changes of the year.[1]

This passage raises a number of points: the colouring of images and the representation of seasonal changes (to which we will return later) but, most important, it establishes the principle that accurate illustration should be essentially a direct 'copy' from the natural object. Pliny talks elsewhere[2] of experience as 'the best teacher' and criticizes the practice of the schools where 'it is more agreeable to sit on benches...than to go out into deserted places and look for different herbs at each season of the year.' The result of

1 Pliny, *Natural History*, Tr. W. H. S. Jones, Loeb Classical Library, 1956, Book 25, iv.

2 Book 26, vi.

this had been a move away from the object-based study propounded by Aristotle (384-22 BC) and his pupil Theophrastus in their botanical writings, and an increasing reliance on authority in place of first-hand experience. This had inevitably led to a degeneration in the accuracy of illustrations which were merely copies of earlier models.

The influential Classical treatises on plants – the most significant being the *De Materia Medica*, a work on the healing properties of plants, by Pliny's contemporary Dioscorides – were lost and have survived only as copies of copies. The most important of these manuscript copies was the *Codex Vindobonensis*, made in Constantinople around AD 512, which contained coloured drawings of plants that are striking in their naturalism – a naturalism alien to Byzantine art of that period. It is possible that the illustrations in the *Codex Vindobonensis* are direct copies from the work of Krateuas of the first century BC, one of the artists cited by Pliny. These copies were themselves repeated in manuscripts produced across Western Europe through the medieval period.

The earliest extant printed herbals in Europe – including the *Ortus Sanitatis* (*c*.1500), and the *Grete Herbal* (1526) – belong to this debased and derivative tradition. Their sources were the degenerate copies of the Greek manuscripts. Their illustrations are crude, generalized to the point of inaccuracy; decorative qualities are emphasized over specifics of identity, with the block-cutter imposing a spurious symmetry wherever possible. Fanciful curlicues and flourishes embellish stems, leaves, tendrils in a random and irrelevant application. Some plants (pl. 32) are distorted by uninformed multiple copying to the point where they are unrecognizable. This copying from earlier sources (and the re-use of printing blocks) without reference to nature led to debased but generally recognizable representations of plants co-existing with fantastic images produced by folk-lore and myth – the narcissus with human heads for flowers in the *Ortus Sanitatis*, the mandrake root in the likeness of the human lower torso and genitals (a version of this is found in almost every early herbal); even Parkinson in 1629 includes in his frontispiece to the *Paradisus* a Scythian lamb (a rooted quadruped!) alongside crude but identifiable images of cyclamen, tulips and lilies. As late as the 1750s Sir John Hill was having old blocks reworked for his *The British Herbal* (1756-7); the craftsman employed to clean the blocks also altered them to disguise their source, cutting out or adding a branch or leaf to each, saying 'I make plants now every day that God never dreamt of.'[3] The true value of such illustrations is no more than that attributed by

6 Chamomile (*Chamaemelum nobile*)

Plate from the *Ortus Sanitatis*, Strasburg, *c*.1500

Woodcut

The *Ortus Sanitatis* was a composition by its printer-publisher Jacob Meydenbach of Mainz with woodcuts based on the small designs in the German *Herbarius* (1485). Here the artist has made an attempt to convey, with the crude and intractable woodcut line, the fine, feathery leaves characteristic of the plant, finishing the stem with a calligrapher's flourish.

3 Quoted in Blanche Henrey, *British Botanical and Horticultural Literature before 1800*, London, 1975, p.94.

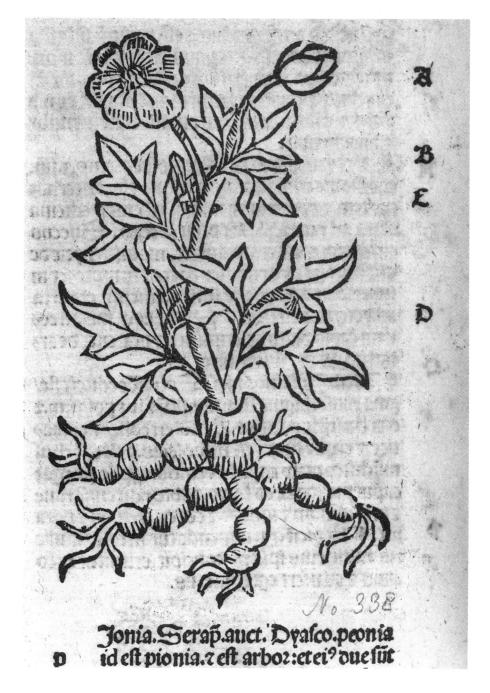

7 Paeony (*Paeonia mascula*)

From the *Ortus Sanitatis*, Strasburg, *c.*1500

Woodcut

The original of this illustration appeared in the German *Herbarius* (1485) where it was a synthesis of the naturalistic and the formulaic (and probably itself a copy of a painted original). This version is severely reduced, the liveliness of the earlier model replaced by stiffness, and described with a crude thickened line. The bud, which in the *Herbarius* was given with reasonable accuracy, has here been entirely misrepresented in the process of copying – no longer round and tight, but elongated like a rosebud or a tulip. The artist of the *Ortus Sanitatis* failed to understand either his model, or the value of a precise copy, and has instead taken parts of the original – the open flower, the roots – and added his own interpretation and embellishment, all of which contribute to a loss of accuracy and sense.

Hieronymus Braunschweig to the pictures in his *The Boke of Distyllacyon* (first published 1500, translated into English 1527): in the conclusion he tells the reader to attend to the text rather than the illustrations 'for the figures are nothing more than a feast for the eyes'.

A major change occurred with the production of Otto Brunfels' *Herbarum Vivae Eicones* (1530-36), 'living portraits of plants'. The text is of little account, heavily dependent on Classical and medieval sources, and often inaccurate. As the title implies, it is the illustrations which are innovatory, and are a landmark in the development of botany as an empirical science. Ironically Brunfels himself dismissed the illustrations as no more than 'dead lines', convinced that they were inferior to the 'right-truthful descriptions' in his text. The artist, despite his skill and originality, is not credited by the author. He was identified as Hans Weiditz, a pupil of Dürer's, only when a cache of the original watercolours was discovered in 1930 in the Botanic Institute at the University of Bern. It has been argued that the value of Weiditz's illustrations is undercut by their being too specific – he presents not the ideal or 'typical' specimen but a particular plant with all its unique blemishes and deformities.

Botanical science of course is founded on the establishment of resemblances (thus Genera and Families) in order to classify the objects of study, whilst at the same time identifying differences (species, varieties). Fundamental was the importance of establishing identity, which could then be used to pattern resemblance or difference. The Brunfels herbal is faithful to its subjects not simply to the level of characterizing the species, but of portraying the unique flawed specimen set before the artist. Weiditz gives us, if you like, a 'warts and all' portrait. The other great influential herbal of the sixteenth century, Leonhart Fuchs' *De Historia Stirpium* (1542), takes the opposite approach; it flatters its subjects, smooths out their imperfections, generalizes and perfects but without distortion, without imaginative embellishment, simply by making whole whatever was damaged, showing as vigorous whatever was wilted. In his preface Fuchs sets out the principles which guided the production of the illustrations:

> As far as concerns the pictures themselves, each of which is positively delineated according to the features and likeness of the living plants, we have taken peculiar care that they should be most perfect, and, moreover, we have devoted the greatest diligence to secure that every plant should be depicted with its own roots, stalks, leaves, flowers, seeds and fruits. Furthermore we have purposely and deliberately avoided the obliteration of the natural form

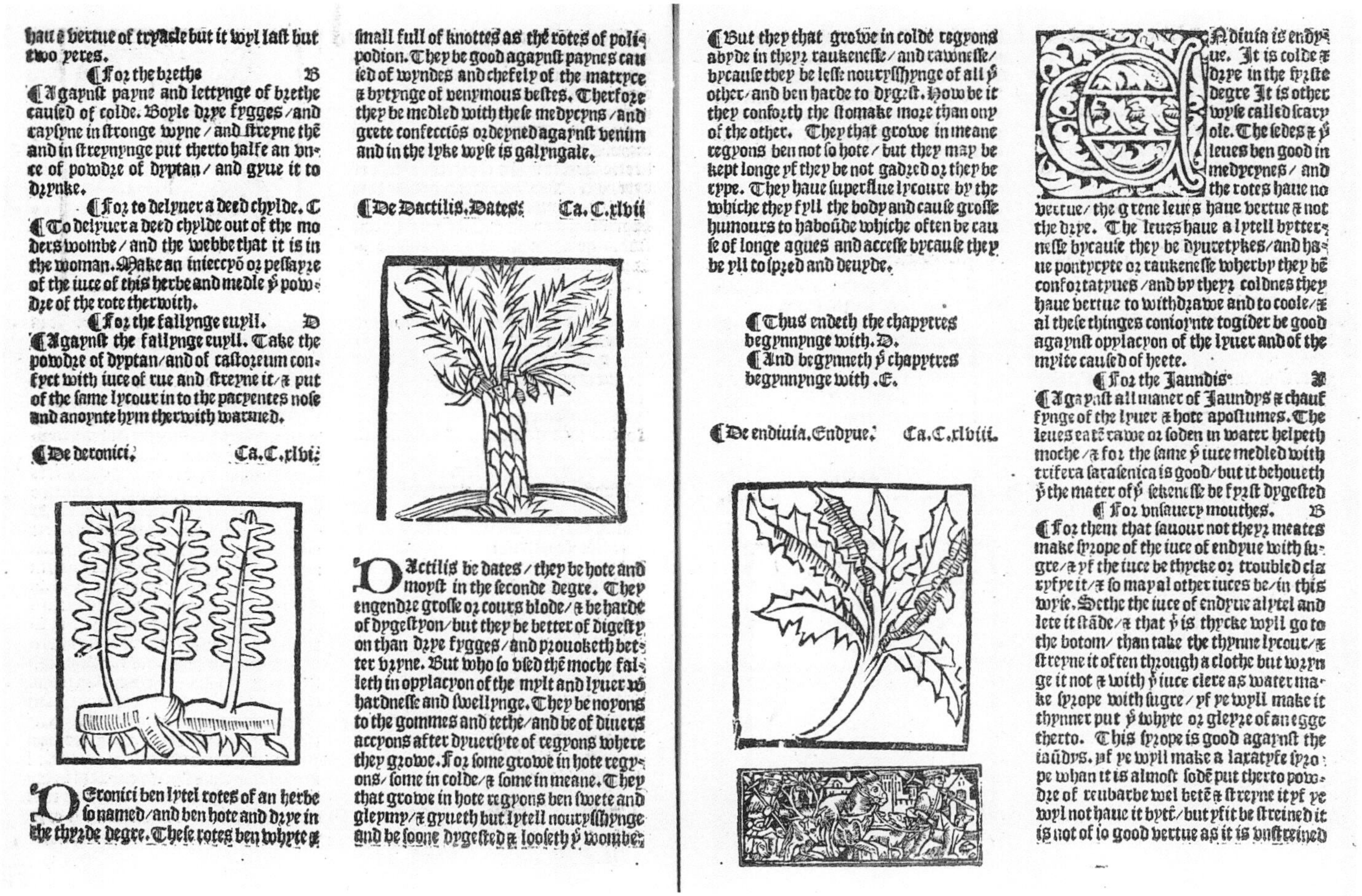

haue ⁊ vertue of tryacle but it wyl last but two yeres.

¶ For the brethe. B

¶ Agaynst payne and lettynge of brethe caused of colde. Boyle drye fygges / and raysyne in stronge wyne / and streyne thē and in streynynge put therto halfe an unce of powdre of dyptan / and gyve it to drynke.

¶ For to delyver a deed chylde. C

¶ To delyver a deed chylde out of the moders wombe / and the webbe that it is in the woman. Make an iniecciō or pessarye of the iuce of this herbe and medle ÿ powdre of the rote therwith.

¶ For the fallynge evyll. D

¶ Agaynst the fallynge evyll. Take the powdre of dyptan / and of castoreum confyct with iuce of rue and streyne it / ⁊ put of the same lycour in to the pacyentes nose and anoynte hym therwith warmed.

¶ De betonici. Ca. C.xlvi.

Betonici ben lytel rotes of an herbe so named / and ben hote and drye in the thyrde degre. These rotes ben whyte ⁊ small full of knottes as thē rotes of polipodion. They be good agaynst paynes caused of wyndes and chefely of the matryce ⁊ bytynge of venymous beestes. Therfore they be medled with these medycyns / and grete confecciōs ordeyned agaynst venim and in the lyke wyse is galyngale.

¶ De Dactilis. Dates. Ca. C.xlvii.

Dactilis be dates / they be hote and moyst in the seconde degre. They engendre grosse or cours blode / ⁊ be harde of dygestyon / but they be better of digestyon than drye fygges / and prouoketh better bryne. But who so used thē moche falleth in opplacyon of the mylt and lyver ⁊ hardnesse and swellynge. They be noyous to the gommes and tethe / and be of dyvers accyons after dyversyte of regyons where they growe. For some growe in hote regyons / some in colde / ⁊ some in meane. They that growe in hote regyons ben swete and gleymy / ⁊ gyveth but lytell nourysshynge and be soone dygested ⁊ looseth ÿ wombe.

¶ But they that growe in colde regyons abyde in theyr raukenesse / and rawnesse / bycause they be lesse nourysshynge of all ÿ other / and ben harde to dygest. How be it they conforth the stomake more than ony of the other. They that growe in meane regyons ben not so hote / but they may be kept longe yf they be not gadred or they be rype. They haue superflue lycoure by the whiche they fyll the body and cause grosse humours to haboūde whiche often be cause of longe agues and accesse bycause they be yll to spred and devyde.

¶ Thus endeth the chapytres begynnynge with. D.
¶ And begynneth ÿ chapytres begynnynge with. E.

¶ De endivia. Endyve. Ca. C.xlviii.

Endivia is endyve. It is colde ⁊ drye in the fyrste degre It is otherwyse called scaryole. The sedes ⁊ ÿ leves ben good in medycynes / and the rotes haue no vertue / the grene leves haue vertue ⁊ not the drye. The leves haue a lytell bytternesse bycause they be dyuretykes / and haue pontycyte or raukenesse wherby they be confortatyves / and by theyr coldnes they haue vertue to withdrawe and to coole / ⁊ al these thinges conioynte togider be good agaynst opplacyon of the lyver and of the mylte caused of heete.

¶ For the Iaundis. A

¶ Agaynst all maner of Iaundys ⁊ chaufynge of the lyver ⁊ hote apostumes. The leves eatē rawe or soden in water helpeth moche / ⁊ for the same ÿ iuce medled with trifera sarasenica is good / but it behoveth ÿ the mater of ÿ sekenesse be fyrst dygested

¶ For unsavery mouthes. B

¶ For them that savour not theyr meates make syrope of the iuce of endyve with sugre / ⁊ yf the iuce be thycke or troubled clarefye it / ⁊ so may al other iuces be / in this wyse. Sethe the iuce of endyve a lytel and lete it stāde / ⁊ that ÿ is thycke wyll go to the botom / than take the thynne lycour / ⁊ streyne it often through a clothe but wrynge it not ⁊ with ÿ iuce clere as water make syrope with sugre / yf ye wyll make it thynner put ÿ whyte or gleyre of an egge therto. This syrope is good agaynst the iaūdys. yf ye wyll make a laxatyfe syrope whan it is almost sodē put therto powdre of reubarbe wel betē ⁊ streyne it yf ye wyl not haue it bytē / but yf it be streined it is not of so good vertue as it is unstreined

8 Double-page spread from the *Grete Herbal*, London, 1526

Woodcut

The *Grete Herbal* is the English version of an anonymous French work, *Le Grant Herbier*, first published *c.*1486-88. The illustrations here are taken from the parent work, from the *Ortus Sanitatis*, and from other sources. They are printed from the same block as the text and the whole is made to look like an illuminated manuscript, down to the inclusion of foliated initials. The plants here – betony(?), date palm and endive – are debased ciphers barely identifiable with their ostensible subjects.

of the plants by shadows and other less necessary things by which the delineators sometimes try to win artistic glory: and we have not allowed the craftsmen so to indulge their whims as to cause the drawings not to correspond accurately to the truth.

It is this method, outlined by Fuchs, of generalizing from nature without digression or invention, that became the dominant mode in botanical illustration. The perfected specimen came to be preferred, or standard, because although the image of a single unique specimen might be literally more truthful, it did not and could not express the larger truth of being characteristic of its species.

Fuchs was responsible for a number of innovations in illustration that were influential, and in some cases became standard practice. He established the principle of showing a plant in flowering and fruiting stages simultaneously (the failure to recognize seasonal changes being one of the faults in Classical botanical illustration cited by Pliny). Many of his plates are composite diagrams combining features of the plant at different seasons, but represented in the one plant as if naturally growing. Though an elegant and economical solution, it is obviously open to misinterpretation, and although other illustrators adopted the device it has been superseded by the equally artificial but less ambiguous separation of the parts. It was also Fuchs who first showed plants life-size, allowing his illustrations both an independence from the text and an equality with it.

The innovations of Fuchs also influenced other compilers of herbals, not by acting as an exemplar of the empiricist approach to illustration but rather as providing a new source of more reliable representations to copy – so we find adaptations of plates from Fuchs in Dodoens (*Cruydeboeck*, 1554, and *Pemptades,* 1583) and in Bock (*Kreuterbuch*, 1546). Direct copying or the modification of printing blocks from earlier publications continued to be the commonest method of illustrating herbals through the sixteenth and into the seventeenth century. The printer Christophe Plantin of Antwerp amassed a collection of blocks which he used as common stock for illustrating the herbals of Dodoens, L'Écluse and L'Obel. John Gerard's *Herball* (1597) used woodcut blocks from the *Eicones Plantarum* of Tabernaemontanus (1590), who had in turn drawn upon Fuchs, Mattioli, Dodoens, L'Écluse and L'Obel. Very few of these books commissioned new illustrations. Gerard, amongst a handful of original plates (sixteen out of 1800), includes the potato, and the milkweed, after a drawing by John White (official artist on Raleigh's expedition to Virginia).

corum, T O M V S Primus. 75

Walwurtz männlin. g 2

9 Hans Weiditz

Comfrey (*Symphytum officinale*)

Plate from Otto Brunfels, *Herbarum Vivae Eicones*, Strassburg, 1530-32

Woodcut

Unlike the artists of earlier herbals, Weiditz did not compromise the naturalism of his plant studies by a misleading distortion of form. Here, in order to include the lower stem and root of his specimen, he has bent it, not by modifying the drawing, but by folding the stem itself and drawing it in that state. This has remained a persistent convention, co-existing with the alternative solution of showing the plant in sections.

In part the arguments for the copying or recycling of existing images were economic; block-cutting was skilled and expensive, as was the commissioning of new drawings and their subsequent translation onto the block. But there was not yet a conscious belief in the value of direct observation over existing images; the belief in established authorities remained strong, as evidenced by the reliance of all the herbals on Classical authors whose work they supplemented but did not seriously challenge. This even-handed approach to originality versus authority is seen very clearly in the watercolour studies of plants by Jacques Le Moyne de Morgues (*c*.1533-88). He seems not to differentiate between those drawings which are plainly taken from nature and those which copy from earlier manuscript sources. Only a certain smoothness, and formality, and a lack of individuality in the specimen distinguish the iris and the violet, both derived from French illuminated manuscripts, from the heart's ease or borage, with the convincing naturalism of their details and imperfections.

* * * * * * *

Most of the early herbalists were engaged in the translation and elucidation of Dioscorides, and the identification of familiar native plants and new introductions with those described in the *De Materia Medica*. Many authors were handicapped in this exercise by the impossibility of matching plants of the Mediterranean regions with those of northern Europe, and with those introduced from other parts of the world. Distortion, misattribution, or omission resulted. Those who recognized the geographical and historical limits on Dioscorides' knowledge were able to take forward the work of studying, describing and picturing new species, establishing botany as an independent discipline. Antonio Musa Brasavola, a physician from Ferrara, observed in his *Examen Omnium Simplicium Medicamentarum* (1536) that 'Not a hundreth part of the herbs existing in the whole world was described by Dioscorides, not a hundreth part part by Theophrastus or by Pliny, but we add more every day.'

Systematic botany – the process of cataloguing all existing species – has its foundations in the herbal tradition with its practical needs for description and discrimination, but there was no 'scientific' method of classification available to the herbalist. Plants in herbals are ordered according to a variety of principles (often within the same book), many of which relate to their medicinal properties. Some followed Dioscorides, Theophrastus and Pliny in distinguishing them according to taste, smell and edibility, or by those parts of the body which they were used to heal. Parkinson's categories

10 Hans Weiditz

Pasque flower (*Pulsatilla vulgaris*)

From Otto Brunfels, *Herbarum Vivae Eicones*, Strasburg, 1530-32

Woodcut

The pasque flower is a seminal image in the history of botanical illustration. It was included by Brunfels only because the artist Weiditz supplied a drawing of it, but it is one of those plants which he had contemptuously designated '*herbae nudae*' because it had no name or description in the Classical texts on which he depended. Subsequently it has come to exemplify the originality and beauty of the work, qualities which rest solely in the illustrations. This is the most commonly reproduced plate from the *Herbarum* and its place in the history of botany was reinforced by its being the type picture of a Linnaean species (as too was the Lady's smock, *Cardamine pratensis* L.) and this despite the criticism of the plates for failing to distil the essence of the species by drawing a synthesis from the study of many specimens.

Within the practical limitations of the woodcut as a reproductive process, the illustrations to Brunfels' herbal achieve an unprecedented degree of naturalistic detail, in this case capturing the distinctive soft, feathery hairiness of the pasque flower. It has also the wilting quality that characterizes so many of Weiditz's subjects, probably because they were not kept in water whilst he drew them.

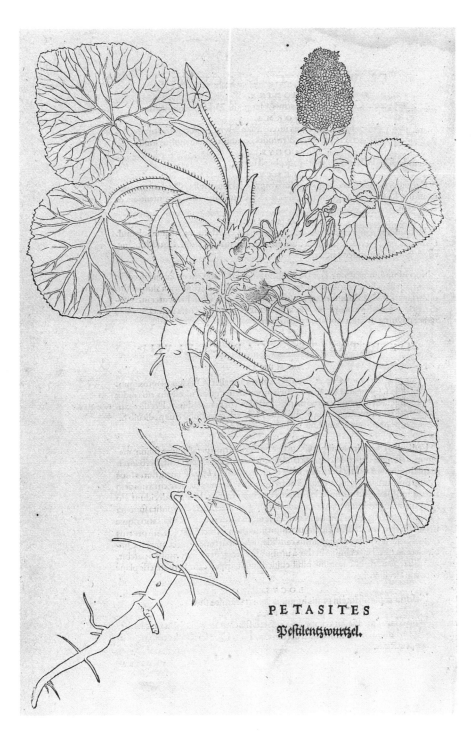

PETASITES
Peſtilentzwurtzel.

11 Butterbur (*Petasites hybridus*)

From Leonhart Fuchs, *De Historia Stirpium*, Basel, 1542

Woodcut

The simple, unshaded outline illustrations in *De Historia Stirpium* suggest that they were designed to be hand-coloured after printing. There are several coloured copies extant, and it appears that the colouring was part of the production process because the colours match so precisely the text descriptions, even to the extent in the 'wegwort' (chicory) (Cambridge University copy) of having both blue and white flowers on the same plant because Fuchs notes that the species is sometimes white-flowered.

In their uncoloured state, the woodcuts have a flat schematic appearance that can be visually confusing where the overlapping parts are unmodulated by shading. To avoid this the specimen is often splayed out so that each part can be seen clearly. The butterbur is a particularly effective example, using the diagonal of the plate to accommodate the long root and broad leaves.

12 Crocus (*Crocus vernus*)

From Leonhart Fuchs, *De Historia Stirpium*, Basel, 1542

Woodcut

One of the innovations in botanical illustration attributable to Fuchs and his artists is the representation, where necessary, of both the flowering phase and a characteristic later stage of a plant's life-cycle. The crocus is a good example in that the flower appears before the foliage and to show either phase in isolation would be seriously misleading. This plate with two figures gives a very complete account of the botanical facts, and does so with the same economy (but greater clarity) than the composite plates where flower and fruit appear simultaneously on a single plant.

The delicate line in these plates marks the highpoint of the woodcut as a medium for scientific illustration. Even so, the woodcut was too crude and inflexible a medium to give the level of detail increasingly required by botanical science, and it was generally superseded by metal engraving by the early seventeenth century.

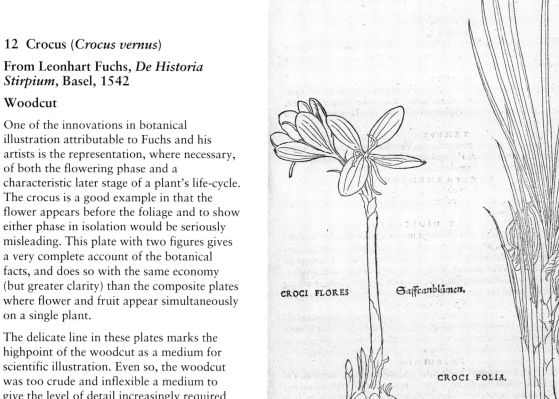

included 'purging', 'venomous', 'sleepy and hurtful', but also 'sweet-smelling' and 'strange and outlandish'; and Bulliard's *Herbier de la France* (1780-95), published after the establishment of a scientific method of classification, nevertheless groups the plants with reference to their properties and medicinal uses into categories that include *plantes suspectes*, *plantes médicinales*, and *plantes veneneuses*. Otherwise plants are ordered alphabetically, by Latin name (*Ortus Sanitatis*, *Grete Herbal*), or in Greek (e.g. Fuchs, who also provided indices in Greek, Latin, German, and by apothecaries' and herbalists' names); by type, as William Coles in *The Art of Simpling* (1656) divided herbs into seven kinds – pot herbs, medical herbs, corn, pulses, flowers, grasses and weeds; or by habitat.

Perceived similarities – functional, visual and structural – were also used as methods of classification, but they were superficial rather than systematic similarities. The lack of a scientific system of classification was felt by Jerome Bock who recognized the corolla, stamens, and pistils as essential parts of the flower, but this perception was not applied until the eighteenth century; nor is it reflected in the illustrations to his *Kreuterbuch* which re-used existing blocks, many of them from Fuchs whose illustrator includes such details as incidentals rather than as essentials. The principle of drawing in detail the different features of the plant was not yet seen to be necessary to botanical illustration, and for all their 'truth to nature' the plates in Fuchs are portraits of character rather than analyses of structure.

Brunfels and Fuchs aside, the illustrations in herbals, whatever their faults and virtues, are both physically and conceptually subordinate to the text. They are supplementary, providing a visual point of reference but not the *raison d'être*. Their value and significance varies with each publication, but generally the illustrations in the earliest printed herbals are little more than decorative motifs breaking up the blocks of text: the *Grete Herbal* (1526) with its images contained within a framing line, is printed to look like a manuscript, including illuminated initials. The strongly symmetrical illustrations resemble printers' devices, decorative motifs, ciphers or pictograms of the plants they 'represent'. With the exception of Fuchs, and some plates in Brunfels, the illustrations in these early herbals are small, cut from the same block as the text (*Grete Herbal*) or fitted into the text in a cramped and rather arbitrary fashion. Visually the text dominates, and the images are peripheral, often literally marginalized; and despite the innovatory significance that distinguished his illustrations, an author such as Brunfels could argue for the supremacy of a derivative and inaccurate text.

13 Chives (*Allium schoenoprasum*)

From Leonhart Fuchs, *De Historia Stirpium*, Basel, 1542

Woodcut

The majority of plants illustrated in Fuchs are perfected types drawn from nature, though some were copied (with the corruption that entails) from Brunfels. This plate shows clearly the characteristic balance in the design, with no part or feature of the plant emphasized at the expense of any other. It is also apparent that the artist has resisted the opportunities for symmetry inherent in the natural form of the plant.

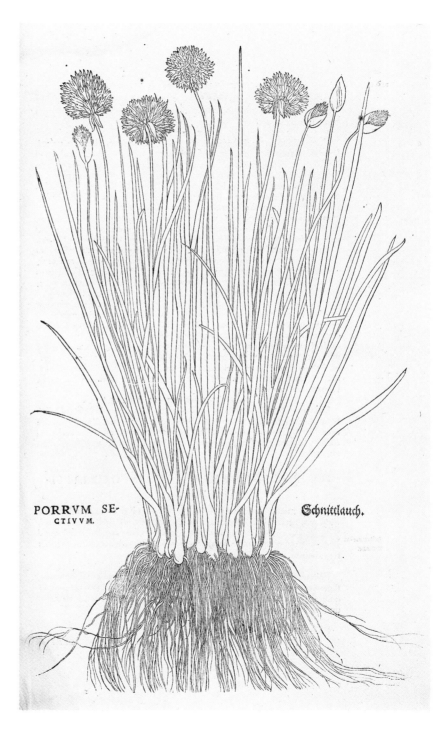

PORRVM SE-
CTIVVM.

Schnittlauch.

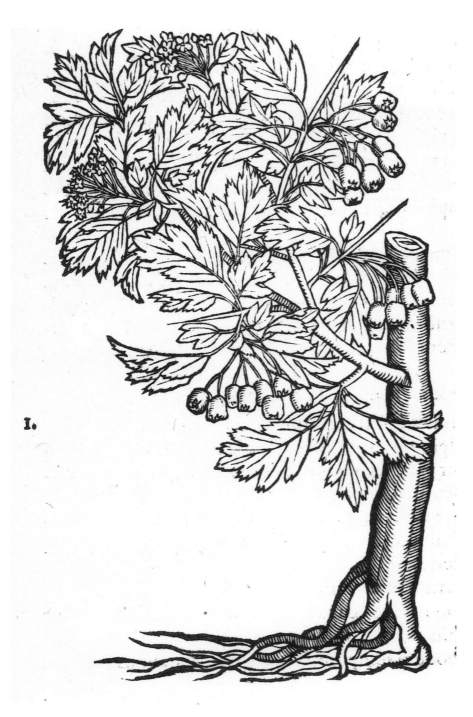

I.

14 Hawthorn (*Crataegus monogyna*)

From Rembert Dodoens, *Stirpium Historiae Pemptades Sex*, Antwerp, C. Plantin, 1583

Woodcut

This plate demonstrates one of the illustrative conventions established by Fuchs and widely adopted in later herbals – the representation of a tree by a rudimentary trunk and a single branch bearing both flowers and fruit. The confines of the plate are also evident in the composition, resulting in the horizontal compression of the roots along the bottom edge, and the leaves and flowers being compressed into the upper-left corner. Despite Fuchs' example of the liberating effects of the page-plate, most herbals reverted to the text figure because it was cheaper, and because many writers were re-using blocks rather than commissioning new illustrations.

15 'Viola purpurea' etc.

From Pier Andrea Mattioli, *Commentarii in Sex Libros Pedacii Dioscoridis*, Frankfurt, 1598 edition

Woodcut

This is a very good example of rectangularization, that is the effect on the draughtsman of working within the confines of a small wood block. Though the artist has clearly observed the natural habit of growth of the '*viola purpurea*', the spreading of the plant is constrained by its confinement in the narrow upright format of the block; in an attempt to compensate for this, several of the flowers and leaves are represented as if spreading into the foreground, but in this linear diagrammatic style they are in fact confused visually with the root system that they overlap.

It is interesting to compare this with the later and more sophisticated representation of a violet, by Bulliard (pl. 23). The illustrator of Mattioli's herbal has not properly understood the construction of the flowers, and so the petals so clearly delineated by Bulliard are here no more than simple overlapping areas of equal size and shape.

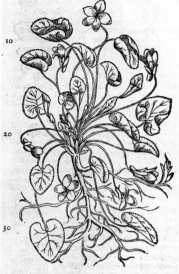

The earliest printed herbals were the product of a culture which was repeating received wisdom, and was not analytical or exploratory. Implicit in the names of the sixteenth-century herbals and plant books is the nature of the knowledge they contain. These are not works of a predominantly empirical science; they contain only what others have observed, believed, understood erroneously or otherwise, and handed down. These books are 'histories', that is they recount an accumulation of plant lore from the past: thus we have Fuchs' *De Historia Stirpium*, L'Obel's *Plantarum seu Stirpium Historia*, Dodoens' *Stirpium Historiae*, Gerard's *Herball or the Generall Historie of Plants* etc. Naturally enough the illustrations too are literally 'received wisdom', copied (and in the process debased and distorted) from earlier models, themselves of varying, often dubious, accuracy. The text itself, a mixture of folk-lore, hearsay, magic and unproven assertion, finds an equivalent in images equally without foundation in empiricism, observation or experience. There is no relationship of interdependence or reference between image and text (even when the correct image is set against the relevant description). The authors rarely refer to what one can see in the illustration itself so the reader must make his or her own connections between the two. Not until other kinds of botanical publication had established the importance of identifying the parts of a plant do we find a herbal such as Elizabeth Blackwell's *Herbarium Blackwellianum* (1754-73) making, albeit at a very basic level, a link between text and images by labelling and distinguishing flower, fruit and seeds.

The obsessive impulse towards symmetry and rectangularization was disrupted by Brunfels and Fuchs: compare for instance the resolute symmetry in Mattioli's illustration of Solomon's seal (pl.16) with the resistance to it in Fuchs' illustration of chives (pl.13), a plant which might equally lend itself to regularity and pattern-making. Other inaccuracies however remained characteristic of herbal illustrations, notably disparities of scale. A fundamental feature of the herbal is the almost unvarying insistence on representing the whole plant as a whole (rather than in parts and sections, as became the established alternative) and thus showing the plant from roots to crown, or alternatively when the root is of no significance characteristically or economically, showing a flower spike. Thus where a plant would in life have extensive roots, or an elongated stem, there must be some modification both of natural habit of growth and of scale to fit it into the block. The bending of a plant or the curling around of the root growth is a common feature of the woodcut herbals. Trees especially are distorted by this process – a square-crowned oak might carry leaves and acorns equal in size

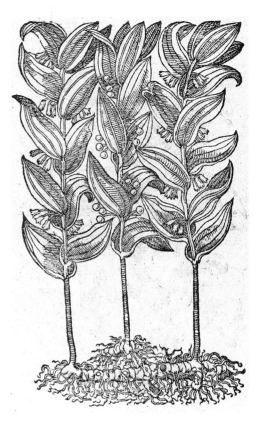

16 Solomon's seal (*Polygonatum multiflorum*)

From Pier Andrea Mattioli, *Commentarii in Sex Libros Pedacii Dioscoridis*, Frankfurt, 1598 edition

Woodcut

Where Fuchs shows this plant as a single arching stem, Mattioli's draughtsman has taken the opportunity afforded by the plant's natural form to create a vertical balanced design.

**17 Lesser celandine
(*Chelidonium minus*)**

**From Pier Andrea Mattioli,
*Commentarii in Sex Libros Pedaccii
Dioscoridis*, Frankfurt, 1598 edition**

Woodcut

This edition of Mattioli's herbal was
enlarged and expanded by the influential
botanist Caspar Bauhin. This plate shows a
number of characteristic features – the
rectangularization and 'all-over' patterning
which result in the image completely filling
the plate, the instinctive impulse to symmetry
and balance in the composition as a whole.
Some hatched shading is employed rather
inconsistently, but generally with the intent to
indicate the undersides of leaves, or those
leaves which are to be read as 'behind' others.
The extent to which the artist has over-
elaborated his subject, whilst offering an
identifiable representation of leaves and
flowers, can be seen by a comparison with
the same plant seen in a contemporary
photograph (pl. 94).

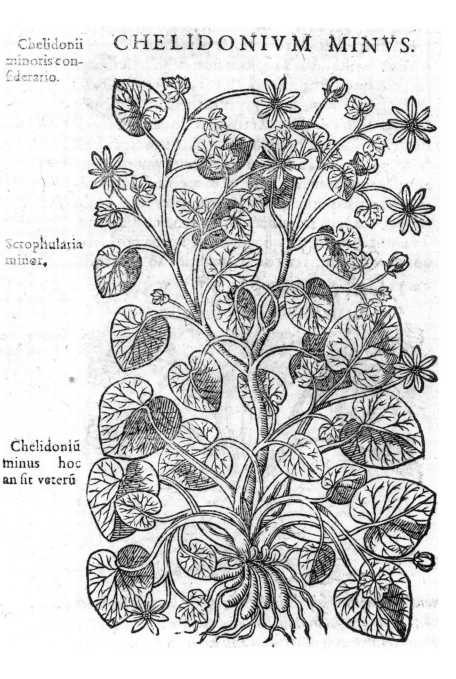

18 Jacques Le Moyne de Morgues (*c*.1533-88)

Borage (*Borago officinalis*)

Watercolour

This is a wonderfully vigorous study, rich with details that give this the convincing individuality of a living plant: the fine hairs on leaves and stems, the precise structure of the flowers, and the torn and discoloured leaf. The whole is approximately life-sized.

Le Moyne's drawings belong to the sixteenth-century revival of naturalism in botanical illustration. They cannot be dated exactly but the watermark in the paper is the same as that used in Paris and Arras in 1568. On this evidence (and the French binding in which they were bound at the time of their acquisition by the Museum) it seems likely that they were made in France before Le Moyne's flight to England as part of the Huguenot exodus in 1572.

19 Chicory (*Cichorium intybus*)

Page from a fifteenth- or sixteenth-century manuscript herbal

Gouache on vellum

This drawing was probably made for a manuscript herbal. It has certainly been bound as part of a volume because the edges of the sheet (and of its six companions) have been gilded. The image is somewhat crudely executed, in part due to the thick dull pigment, but the subject is readily identifiable. Though the artist lacked the skill to deal convincingly with representational devices such as foreshortening – see the fifth leaf above the root – the actual structure of the plant suggests that he has studied either a living specimen or a botanically-accurate model. The branching of the stems, the size and placing of the bracts and buds, are all correctly given.

to the width of the trunk; in Dodoens' herbal a favourite manner for representing trees and shrubs is to show the roots and a small length of trunk to one side of the block, leaving space for a characteristic branch to fill the diagonally opposite corner (see pl. 14) Whilst this is a little more convincing, it still seriously misrepresents the relative sizes of the different parts, and suggests an audience for these images which was sufficiently sophisticated not to interpret them literally. The use of small blocks invariably led the artist to work right to the edges of the block; the result is the kind of illustration typified by Mattioli which is a dense configuration of linear pattern with a powerful tendency to symmetry (pls. 16, 17). Habit of growth was also compromised by the confining character of an upright oblong block – the violet and the cyclamen in Mattioli are obvious examples of low-growing spreading plants coerced into a wholly inaccurate upright habit. Compare for instance Mattioli's cyclamen with the more naturalistic example from the *Hortus Floridus* (1614, pl. 34). Though published as a florilegium, this ostensibly decorative volume has more botanical authority than the superbly decorative images from Mattioli's herbal. Admittedly the *Hortus Floridus* was published 70 years after the first appearance of Mattioli's *Commentarii*, but the examples illustrated are from the 1598 edition. It is the technical innovation of copper-plate engraving which has liberated the artist of the *Hortus Floridus* as much as the shift towards direct observation as the premise for botanical illustration.

In Fuchs most of the plants were shown life-size where possible, and the image begins to assert a physical and intellectual equality with the text. *De Historia Stirpium* inaugurated the 'page-plate' where each figure is given a whole page to itself, and is rendered physically independent of the text, both by its separation and by its authority, founded in its 'truth to nature'. Though as we have seen, later authors reverted to text figures, Fuchs marks the beginning of a new emancipation of the image, and its establishment as a primary source for botanical data. Fuchs, and Brunfels too, also originated pictorial devices to show the whole plant without distortions of scale, for instance by exploiting the diagonal of the plate as in Fuchs' butterbur (pl. 11) or by bending the stem – literally rather than artistically – as in Brunfels' comfrey (pl. 9). Such devices survive into the graphic lexicon of contemporary illustration.

Though isolated instances of herbals – 'plants cultivated for use' (Sir John Hill's *The British Herbal*, 1756-7) or those 'now used in the practice of Physic' (Elizabeth Blackwell, *Herbarium Blackwellianum*, 1754-73) –

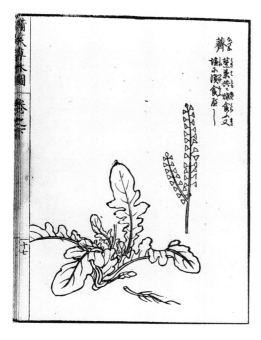

20 Kitago Shimei

Shepherd's purse
(*Capsella bursa-pastoris*)

Plate from Takebe Seian, *Biku Somoku*, 1833

Woodcut

This is a 'famine herbal' describing and illustrating wild plants that could be used for food in times of scarcity. The images were probably no more than basic identification guides for a rural population already familiar with the plants that they described.

appear in the seventeenth and eighteenth centuries, they are primarily a product of the sixteenth- and early seventeenth-century view of plants as no more than sources of food and medicine. The divorce of botany from medicine and its establishment as a separate discipline was fostered by the arrival in Europe of new species which subverted the authority of Pliny, Pseudo-Apuleius and Dioscorides, and by an approach to science which was posited on exploration, experiment and observation. New methods of classification were founded on the structure of the plant itself, independent of its uses and virtues, its habitat or its definition in classical sources. Botanical illustration is given a new value in the advance of knowledge, and becomes a tool for the botanist, the taxonomist, the gardener and the collector.

The progress of Japanese plant lore and botanical study followed a similar pattern to that we have observed in Europe. A close dependence on authorities from another age and culture was characteristic of Japanese botany until the early seventeenth century. As Classical sources were to the European herbalists so Chinese and Korean *materia medica* were to the Japanese.[4] Nor did the Japanese possess a real sense of botany as an independent discipline; rather it was subsumed into natural history, and like the European herbal of the fifteenth and sixteenth centuries the Japanese *honzo* (herbal) was encyclopaedic and might include animals, geological material, and inorganic substances in addition to plants.

European writers had struggled to match their native flora with the descriptions found in Classical authors; the same exercise of identifying native species with those described in earlier foreign – usually Chinese – sources took place in Japan. Chinese and Korean works were long regarded as authoritative, were extensively studied, and were published in Japanese editions. The herbal – a work devoted to an account of utilitarian plants – survived much longer in Japan than in Europe: a so-called 'famine herbal', *Biku Somoku Zu, Kan, Kon* [Pictures of plants necessary in preparation for famine relief] by Takebe Seian, first published in 1771, was reprinted as late as 1833. As with so much else in Japanese botany, the tradition of the famine herbal itself had been adopted from China.

The catalyst to this process, and to the recognition of botany as an independent subject of study, was in part the influence of Western science, which filtered through only very gradually via the Dutch trading post in Nagasaki, during the period of Japan's self-imposed isolation from 1639 to 1858.

4 Literally 'substances for curing or healing' but generally used to describe the literature pertaining to these things.

the inside of the admirable

21 Jacques Le Moyne de Morgues (*c*.1533-88)

Sweet violet (*Viola odorata*) and red admiral butterfly (*Vanessa atalanta*)

Watercolour

The source of this watercolour has been identified as a French illuminated manuscript (B.L.Add.MS.35214, f.50r.) but there is plenty of internal evidence to show that the plant was not drawn from life. It is stiff, formal and stylized, with the leaves arranged in a near symmetrical balance. Both leaves and flowers are flat and lacking in detail – there are no veins, no blemishes, no irregularities. The colours too have a parched brownish tone quite distinct from the freshness of Le Moyne's study of borage (pl. 18) for example. By contrast the butterfly (seen from beneath) is a fine naturalistic study.

22 Wild Strawberry (*Fragaria vesca*)

Plate 77 from Elizabeth Blackwell's *Herbarium Blackwellianum*, Volume I, Nuremberg, 1754-73

Colour engraving

The *Herbarium* was a reworking of Blackwell's earlier publication, *A Curious Herbal* (1737) which had been undertaken to redeem her husband from a debtors' prison. She was encouraged in the enterprise by Sir Hans Sloane, and took lodgings near the Chelsea Physic Garden in order to have her subjects to hand. She drew, engraved and coloured the plates herself, and the publication was a commercial success. It was re-published and expanded by Ehret's patron, C. J. Trew, who employed Nikolaus-Friedrich Eisenberger (1707-71) to revise the first 500 plates and to add a sixth volume.

Blackwell's originals were botanically adequate, and included keyed details and simple floral dissections, but were somewhat naive in execution. Eisenberger's contribution was to produce something artistically superior. He was himself a competent botanical illustrator.

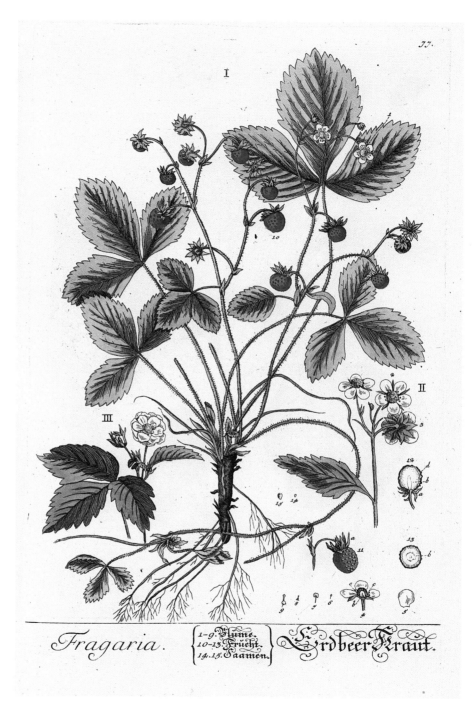

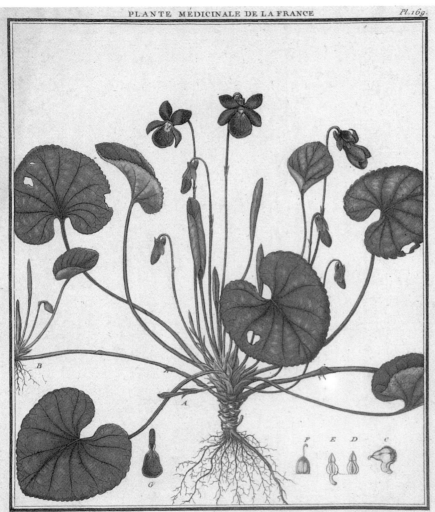

Pl. 169.

LA VIOLETTE ODORANTE. FL. FR.

Viola odorata *L. S. P. Syng. monogam. 1324. on trouve communément cette plante en fleur en mars et avril, dans les prés, les bois, les haies, les jardins, elle est vivace: elle n'a point de tige, chaque fleur a pour support un pédoncule stipulé qui part immédiatement du collet de la racine... ses fleurs sont composées d'un calice de cinq feuilles, d'un corolle de cinq pétales inégaux, le pétale inférieur, fig. G. est le plus remarquable, il se termine postérieurement en un nectaire en éperon; au centre des pétales on trouve cinq étamines réunies à leur base, et rapprochées chacune autour d'un style recourbé... pour fruit elles ont une capsule uniloculaire et trivalve... ses feuilles sont cordiformes, soutenues chacune par un pétiole grêle qui part du collet de la racine ainsique plusieurs dragons A, B, au moyen desquels cette plante se multiplie prodigieusement.*
N° B les fig. C, D, E, F, sont dessinées à la loupe, la fig. C représente les étamines entourant le pistil... la fig. D. est celle d'une étamine séparée... la fig. E. celle d'une des deux étamines dont l'appendice se prolonge dans le nectaire de la fleur... la fig. F. représente le germe et son style.
Cette plante est très connue tant par le fréquent usage qu'on en fait en médecine, que par l'odeur agréable que répendent ses fleurs: nous parlerons de ses propriétés dans le DISCOURS sur les plantes médicinales.

23 Pierre Bulliard

Violet (*Viola odorata*)

Plate 169 from *Herbier de la France* Paris, 1780-95

Colour engraving

All the 600-odd plates in this ambitious herbal were drawn, engraved and colour-printed by the author. Functionally the work belongs to the herbal tradition for it illustrates and orders plants according to the same divisions used by the sixteenth-century herbalists – medicinal, poisonous, edible – but it is also botanically valid since the plants are classified according to the Linnaean system.

Bulliard had a strong sense of design and this violet is beautifully placed within the borders of the plate, its habit of growth described with elegance and accuracy. The holed leaves suggest that the artist has, in the manner of Weiditz for Brunfels, drawn a particular rather than a generalized specimen. In fact such damage is characteristic of the violet, and must be seen as a typical though accidental and random feature and not unique to a single specimen.

The delicately-modulated colour is achieved entirely by printing with three tint plates over an engraved black outline and shading. The *Herbier* is one of the most impressive examples of colour-printing in the history of botanical illustration, and is unusual in needing no supplementary hand-colouring.

Florilegia and Pattern Books

The word 'florilegium' (literally 'flower book') describes a category of books where the plates are intentionally more significant than the text; indeed the text may amount to no more than captions. The florilegium was a characteristically seventeenth-century production, with four important examples published between 1608 and 1616. Manuscript florilegia were also being painted for wealthy patrons. This was an age when flowers were increasingly grown purely for their decorative qualities, their colours and forms, and not as previously for their practical use in medicine and cookery. The kitchen garden, the herb garden and the physic garden were now joined by the flower garden – the 'garden of earthly delights' – and it became fashionable to collect and cultivate flowering plants. Indeed plants and gardens became status symbols, indicative of wealth and of intellectual engagement. The process had already begun with the herbals published towards the end of the previous century. Gerard (1597) for instance, includes the fritillary (*Fritillaria meleagris*) not for its virtues but for its beauty alone; as he explains 'Of the faculties of these pleasant flowers there is nothing set down in the ancient or later writers but [they] are greatly esteemed for the beautifieng of our gardens, and the bosomes of the beautifull'. It was also of course a time when new species were appearing in Europe as a consequence of travel and trade, and when horticulture saw the development of new varieties, hybrids and cultivars – the cultivation of the tulip being perhaps the most famous example of the current obsession with the new.

Florilegia were in a sense catalogues of collections, for a garden (certainly in the seventeenth century) was a collection of ephemeral exhibits, beautiful, often rare and exotic, imported or otherwise acquired at great cost. Some florilegia were records of specific gardens. The *Hortus Eystettensis* (1613) by Basil Besler (1561-1629), a Nuremberg apothecary and botanist, is the earliest pictorial record of flowers in a single garden – the garden of his patron, Johann Konrad von Gemmingen, Prince Bishop of Eichstatt. So too the manuscript florilegium compiled between *c.*1651 and 1670 by Johann Jakob Walther which records the plants in the garden at Idstein (near Frankfurt) created by the Count of Nassau. This was not published but exists as two volumes of watercolours (133 flower studies plus views of the gardens) in eighteenth-century bindings.[1]

1 There is a second version in the Bibliothèque National in Paris which has very similar but not identical plates dated 1652-65; a third copy consisting of 200 plates, formerly in Darmstadt Landesbibliothek, was destroyed during the Second World War.

24 Johann Jakob Walther (*c*.1604-77)

Cowslip (*Primula veris*), Bird's eye primrose (*Primula farinosa*), wild hyacinth (*Endymion Nonscriptus*), double primrose (*Primula vulgaris*), polyanthus (*Primula* x *variabilis*), narcissus (*Narcissus* sp.)

Plate from the Nassau Florilegium, *c*.1650-70

Watercolour

25 Johann Jakob Walther

Paeonies (*Paeonia officinalis*)

Plate from the Nassau Florilegium, *c*.1650-70

Watercolour

There is a naive formalism to Walther's illustrations which supersedes and often precludes naturalism. The stiffness of his plant portraits mirrors the formal plantings of seventeenth-century gardens, with each plant 'displayed', set at a distance from its neighbours. Plants naturally single-stemmed and erect in habit were preferred – tulips, crown imperials, narcissi, hyacinths, foxgloves, lilies. In Walther's illustrations even those of a spreading or bushy growth such as paeonies and roses have imposed upon them an upright posture, so that each flower can be shown in isolation. The paeony is supported here on impossibly slender stems elegantly entwined so that the composition accommodates and balances each flower.

The florilegium was the fruit of several convergent influences, such as the fashion for floral motifs as decoration in the applied arts, and the collecting and cataloguing impulse which underlay so many of the scientific developments of the seventeenth century. Most immediately, the production of florilegia was a response to the wealth of new plants, and in particular the colourful showy exotics, that were arriving in Europe from the southern and eastern Mediterranean regions, north Africa and the Americas, from the second half of the sixteenth century. These new plants were brought in with other curiosities, both natural and man-made, by voyages of discovery and through trading contacts. It is perhaps significant that some of the most important of the early botanic gardens were established in or near major trade centres, such as Amsterdam and Venice (the botanic garden at Padua). With so many plants arriving in northern Europe, where they were enthusiastically traded, swapped, cultivated and hybridized, the gardens of the wealthy collectors of the period became living extensions of the *Kunst* – and *Wunderkammern*, collections of curious and wonderful artificialia and naturalia. Indeed contemporary commentators made no distinction between plants and other rarities – all were subsumed in the term 'curiosities': the Bishop of Eichstatt writing to Philip II of Pomerania, says 'as to where the *rara* and curiosities mentioned in your letter can be acquired, I can tell you nothing more than that the garden plants were brought back through the offices of local merchants, above all from the Netherlands...'.[2] Emanuel Sweert, trader in all kinds of curiosities including plants, published a *Florilegium* that was both picture book and sale catalogue in 1612.

2 Quoted in Nicholas Barker, *The Hortus Eystettensis: The Bishop's Garden and Besler's Magnificent Book*, London, 1994, p.1.

Kunst- and *Wunderkammern* are extraordinarily various in their contents. Collecting was rarely thematic or 'scientific' in any sense; rather it was voracious and seemingly indiscriminate, the only criteria being that the objects chosen were in some way rare, extraordinary, new or curious. These collections were amassed by princes and by virtuosi: John Evelyn was the possessor of a famous cabinet of curiosities (he was also a writer on natural history subjects e.g. *Sylva*, his treatise on trees); and the plant-collecting John Tradescant, who founded a botanic garden in Lambeth in 1629 and, with his son, also amassed a collection of rarities which became known as 'the Ark'.[3] The flower pieces and still lifes by Dutch painter Rachel Ruysch (1664-1750) were inspired by, and often depicted, the collection of shells, fossils, minerals and rare plants owned by her father, scientist Anthony Frederick Ruysch. Most of these collections contained natural objects (naturalia), not the 'most plain and pure' but rather examples of *Nature Erring, or Varying...Nature Altered* or wrought'.[4] Similarly the creator of a garden

3 Augmented by the collection of the late Elias Ashmole it was the foundation of the Ashmolean Museum, Oxford.

4 Francis Bacon, *Advancement of Learning*, II.8.

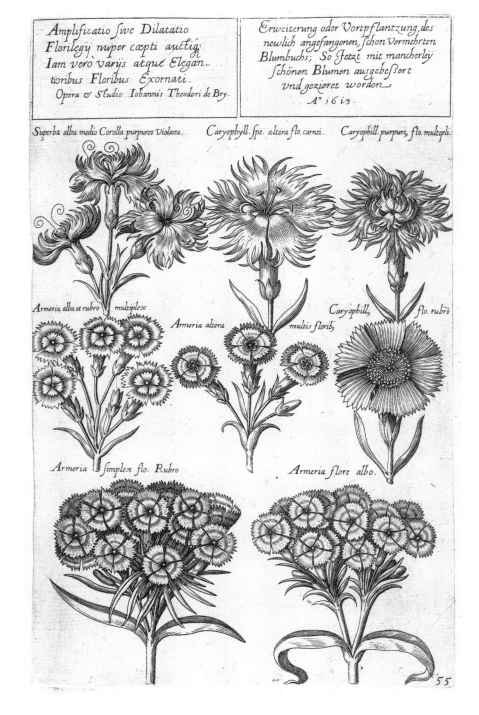

26 Carnations (*Dianthus caryophyllus*) and sweet williams (*Dianthus barbatus*)

Plate from J. T. de Bry, *Florilegium Novum*, Oppenheim, 1612

Engraving

As its title implies, the *Florilegium Novum* was concerned with representations of new garden flowers. Generally described as a pattern book, De Bry's *Florilegium*, though it borrows plates from Vallet's *Le Jardin du Roy*, presents many of its subjects more in the manner of a trade catalogue. Indeed it shares many of its plates with Emanuel Sweert's *Florilegium*, a catalogue of plants offered for sale, which was published the same year.

27 Johann Jakob Walther (*c.*1604-77)
Carnations (*Dianthus caryophyllus*)
Signed *Johan Walter* and dated *1654*
Watercolour and bodycolour

28 Johann Jakob Walther

Tulips (*Tulipa* sp.)

Watercolour and bodycolour

Both plates, from the Nassau Florilegium,
demonstrate Walther's skill in decorative
composition. Though both groups are
evidently 'arranged' to present a formal
bouquet-style grouping, the natural habit of
the plant is approximated. The carnations are
shown growing in a pot, which was the usual
method of cultivation at this time.

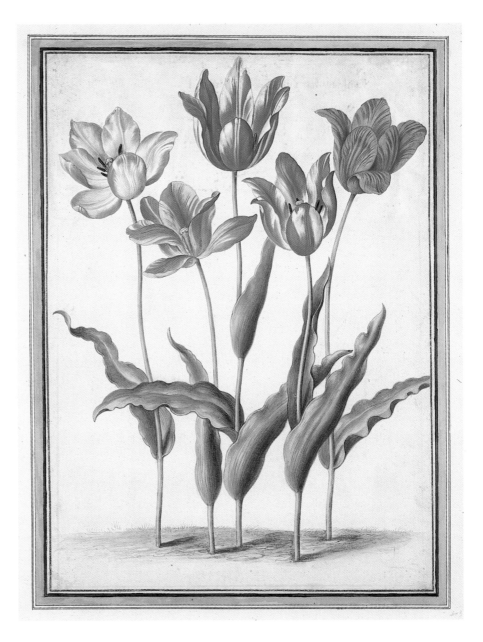

was motivated to acquire the new: plants imported from exotic newly-discovered and colonized lands, or plants that were exceptional for their flowers, fruit or foliage. Thus florilegia tend to show several varieties or colour variations of the same flower – the narcissus, anemone, carnation, and above all the tulip, a plant valued for its unpredictable tendency to 'break', producing bizarre new colour combinations. (Caused, as we now know, by a virus, these broken tulips were notoriously unstable so an illustration might exist as the only evidence of a particular colour form.) Specialist florilegia were devoted to tulips alone or to bulbous plants including lilies, daffodils and narcissi. The *Livre de Fleurs* (1620) by Langlois, self-described as an *enlumineur* (an illuminator of manuscripts), was probably intended in the first instance as a pattern book for his trade but it sets out 23 named varieties of tulip, together with irises and narcissi. The *Hortus Floridus* (1614) by Crispin de Passe the Younger was the first published instance of a broken tulip; numerous examples appear in the *Hortus Eystettensis*, and later in the Walther florilegium.

Florilegia illustrate exclusively cultivated plants, mostly ornamental and valued for the decorative character of their flowers, fruit or foliage. The composition of the plates, with plants grouped for decorative effect, reflects this. Florilegia show us perfected portraits of individual plants, without repetition. Though several kinds of paeonies, tulips or narcissi are represented, each is a different variety, distinguished by colour or form (thus single and double forms of narcissus, and pink and crimson paeonies in the Walther). More especially with specimens such as flowering plants, where the flowers themselves are the most transient part of an essentially ephemeral organism, the catalogue preserved a record of what would otherwise pass away. The third edition of the *Hortus Eystettensis* (1713) has the word *olim* (meaning 'once' or 'formerly') in the title, indicating that the original garden no longer existed, and thus the book, in its various editions, was the only monument to its magnificence. Likewise, the Idstein garden had all but disappeared by 1795, but is now undergoing restoration using Walther's illustrations as evidence for the reconstruction. No written record of the Idstein garden would convey so powerfully the profusion, variety and colours of the plants it contained. Thus the florilegium stands in for the thing itself, becomes a permanent portable substitute for the impermanent and fixed, a garden and its plants. Other catalogues of collections – geological specimens, shells, gem stones – are less dependent on illustration, and may exist simply as elaborated lists, because the objects described are not subject to change or decay. Such catalogues would include Imperato's

Historia Naturale di Ferrante Imperato (Naples, 1599) which has only a frontispiece showing the cabinet of curiosities and Nehemiah Grew's *Museum Reglis Societatis...or a catalogue and description of the artificial and natural rarities belonging to the Royal Society* (1681), which has only a few illustrations of selected items from the many described in the text.

The picturing of plants in the form of a florilegium, which has no scientific purpose, no intent to analyze, classify or otherwise explore its subjects, no text, and no argument, is primarily a statement of possession, of ownership. By the eighteenth century 'curiosity' and collecting were suspect activities, and collections predicated on beauty or curiosity value alone were seen as indulgent, even corrupt. Lord Kames[5] talks of the 'love of novelty' which 'prevails in men of shallow understanding' and condemns as 'deficient in taste' those who prefer 'things odd, rare or singular, in order to distinguish themselves from others.' In the eighteenth century distinctions were drawn between the decorative botanical image, and the purposeful analytic image. For example, in order to legitimize collecting as a scientific endeavour rather than a licentious acquisitiveness, the engraved version of the Benjamin West portrait of Joseph Banks shows, besides an impressive array of ethnographic artefacts, an open portfolio of botanical drawings. The strictly scientific image of the plant asserts the validity of Banks' appropriation of material from other cultures. Just as the plant, described, analyzed and classified, will advance knowledge, so the other objects are not to be seen as booty, but as specimens to be used in similar ways.[6]

The collection of plants, as indeed of other categories of exotica, was contingent upon wealth and leisure, and was motivated by curiosity, novelty, exoticism and rarity. Those who established notable gardens were royalty and aristocracy, merchants, bishops, people with independent incomes. The act of collecting was part of the commercial exchange with, and exploitation of, other cultures. It is not surprising therefore that the collection, study and depiction of plants in the sixteenth, seventeenth and eighteenth centuries was focused on Holland, Germany, northern France, and England – the great centres of trade and colonial power. Sometimes plants were traded as a commodity – as with the tulip during the great speculative mania of the 1630s – or they themselves became the currency of exchange between botanists and collectors. Images were also exchanged, or produced in one locale and sent to another to be reproduced in books, as in the eighteenth century many of Ehret's works were made in England and sent to Trew for publication in Nuremberg.

5 In *Elements of Criticism*, 1762, 6th edn, 1785, I, p.269.

6 See Thomas, N. 'Licensed Curiosity: Cook's Pacific Voyages' in Elsner, J. and Cardinal, R. eds, *The Cultures of Collecting*, London, 1994, pp.116-36.

29 Jacques Le Moyne de Morgues (*c.*1533-88)

Wild strawberry (*Fragaria vesca*) and female emperor moth (*Saturnia pavonia*)

Watercolour

The form of this plant follows closely, though not exactly, the illustration in Fuchs' *De Historia Stirpium* (1542), p.853. Le Moyne's botanical work is part of a complex network of copying . He used Fuchs, or herbals derived from Fuchs, as sources for several other studies including the lily of the valley and the cyclamen (both in the V&A). He reworked the subject himself in another watercolour, a more formal composition minus the moth, now in the British Museum, and in the much-simplified woodcut in his pattern book *La Clef des Champs* (1586). The V&A version is almost certainly the source for plate 103 in the *Altera Pars* of the *Hortus Floridus* (1614). The composition combines naturalistic detail with a contrived formality.

30 Basil Besler (1561-1629)

Cyclamen purpurascens and *Cyclamen hederifolium*

From *Hortus Eystettensis*, Volume II, 1613

Engraving

This magnificent florilegium was one of the first botanical books illustrated with engraved plates. Besler worked intermittently for sixteen years on the drawings, which were engraved by Wolfgang Kilian (1581-1662), and others, as 374 engraved plates illustrating more than a thousand flowering plants from the gardens of the Prince Bishop of Eichstatt. As this example demonstrates, the plates are strikingly decorative, and though there are no botanical details the habit and character of the plants are clearly shown without sacrificing the balance of the design or the evident delight in pattern and symmetry.

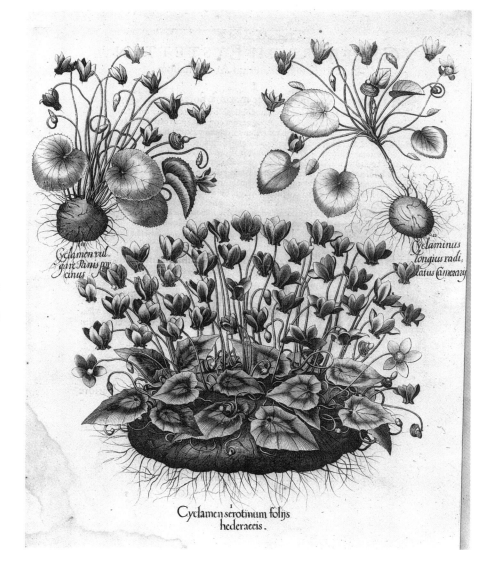

The characteristic method of classification in a florilegium is arrangement of plants by season. The *Hortus Eystettensis* is divided into Spring, Summer, Autumn and Winter sections; the *Hortus Floridus* is 'Divided in the four seasons of the year'; and the plants in Walther are ordered by season beginning with the winter-flowering plants such as the snowdrop (*Galanthus nivalis*) and the Christmas rose (*Helleborus niger*) and passing through a succession of now-familiar garden plants, many of them then relatively new to northern Europe – narcissi, tulips, paeonies, roses, aquilegia, carnations, foxgloves, the marvel of Peru, sunflowers – and ending with compositions of fruit flowers, and insects arranged like still lifes or the marginalia of illuminated manuscripts.

Walther's plant portraits are deficient in botanical knowledge; though the flowers themselves are well-observed, almost every plant, whatever its natural habit of growth, is reduced to a single stem standing stiff and straight (pl. 24), or given a wholly spurious sinuosity to enhance the composition (pl. 25). Many are shown with appropriate seasonal companions, and most are shown as if growing. This singling-out of specimens, and setting them out with plenty of space between and around them, is not merely a compositional device, but actually reflects the characteristic planting plan of seventeenth-century gardens. The plants were set out in box-edged flower beds or parterres; a contrast of colours and forms was preferred, so many different species were mixed together in what the Victorians called a 'promiscuous' bed, the taller plants such as hollyhocks at the back and low-growing plants such as primulas at the front. The plants were kept tidy and not allowed to trail or over-grow their neighbours; they were well spaced and planted with mathematical regularity with plenty of bare earth showing between them. Such plantings are shown clearly in the frontispiece to de Passe's *Hortus Floridus* and in Walther's own bird's-eye view of the Idstein garden which prefaces his florilegium. Many of the illustrations in both Walther and de Passe show the plants actually set in the soil, the latter often from an artificially low viewpoint. De Passe even composed his plates, most of them 'landscape' format, as Dutch landscape scenes in miniature. Many of the plates set a low viewpoint so that one is looking up at plants which loom above a flat horizon and are silhouetted against the blank page as if against the sky; in effect the viewer shares the perspective of the mouse that de Passe introduces into one of his plates of crocuses. Though produced as a decorative pictorial record of a garden, real or imagined[7], the engravings are masterpieces of horticultural illustration, which accurately observe the structural details and habits of growth of their subjects.

7 The frontispiece to the 'Aetas' section shows a garden not unlike that at Idstein illustrated in Walther's Nassau *Florilegium*, but it has not been identified.

31 Basil Besler

Cuckoo-pint (*Arum maculatum*) (I), Arum sp.(II), spring meadow saffron (*Bulbocodium vernum*) (III) and *Hyacinthus* sp.(IIII)

Plate from *Hortus Eystettensis*, 1613

Engraving

The *Hortus Eystettensis* is one of the most ambitious florilegia ever produced. The plates are notable for their elegant design and decorative *mise-en-page*. We know from the copious correspondence (see Nicholas Barker, *The Hortus Eystettensis: The Bishop's Garden and Besler's Magnificent Book*, British Library Publications, 1994) surrounding its production that it was designed to be coloured, or 'illuminated' and a number of variant hand-coloured copies survive. However the quality of the engraving is such that even a 'white' (i.e. plain) copy has great aesthetic value, and is indeed preferable to a poorly-coloured edition where the colours are botanically inaccurate, or are applied so crudely that they disguise the finer details.

The *Hortus Floridus* itself was a florilegium with a strong horticultural bias (later editions append a treatise on tulip growing and add thirteen more tulip plates to the Spring section, some of which show the hoops and stakes used to support the plants in full bloom); it also had close affinities to the naturalistic landscape and still life traditions in Dutch art. The *Hortus Floridus* is generally bound with a volume known as the *Altera Pars*; the former is the work of Crispin de Passe the Younger (1597/8-post 1670), the latter plausibly attributed to his father, also Crispin de Passe (the inscription at the base of the engraved frontispiece – missing from V&A copy – runs 'Formulis Crispiani Passaei et Joannis Walnelij'). There is a marked difference in style between the two volumes, which supports the attribution of the *Altera Pars* to another hand: the *Hortus Floridus* is devoted exclusively to cultivated ornamental plants which are depicted with a high degree of naturalism; the *Altera Pars* is more naive in its representations – the plants are depicted in a flatter, more frontal manner, and are more consciously decorative and stylized. Mostly we are shown only the flowering or fruiting stem, rather than the whole plant as in the *Hortus Floridus*. The plants are presented in unrelated pairs, 120 numbered species in all. A variety of styles are evidence of the images having been culled from different sources, notably Adriaen Collaert's *Florilegium* (lavender, currants), and the watercolours of Jacques Le Moyne de Morgues – but not, as is often claimed, from his woodcuts in *La Clef des Champs* which offer no direct match with the images in the *Altera Pars*, and are in any case too crudely simplified to serve as models for the detailed, if mannered, *Altera Pars* engravings. (Amongst Le Moyne's watercolours in the British Museum and the V&A are several examples that have exact counterparts in de Passe's engravings: the common mallow appears in reverse as no. 11, and a modified version of the strawberry is no. 103. Interspersed amongst the named and numbered species are fanciful decorative plant motifs (*rinceaux*), together with animals and insects which seem to have escaped from some illuminated manuscript. The plants themselves are mostly native wild plants and trees, edible and medicinal. The *Altera Pars* thus looks back to the herbal tradition in both subject and style, but has adopted it for an unashamedly decorative purpose.

It is often hard to distinguish between florilegia which are designed as records of garden plants, and those which are first and foremost pattern books, since the standards of naturalism and botanical draughtsmanship are more or less the same. If anything, the pattern books tend to a greater degree of pictorial naturalism, and the florilegia to a formality of form and

32 'Anfodilo'

Plate from G. Colombina,
***Il Bomprovifaccia*, Padua, 1621**

Woodcut

The cumulative debasement of herbal imagery is evident here. The form of the main flower is ambiguous, with the shading confusing rather than clarifying. It is the result of the ignorant copying of an imperfectly-understood original. Columbina's book is a very late example of the re-using of woodblocks for herbal illustration; many of his images first appeared in the Latin *Herbarius* of 1484.

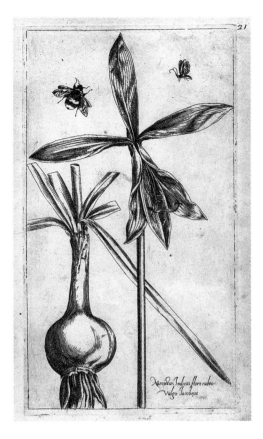

33 Pierre Vallet (b.1575)

Jacobean lily (*Sprekelia formosissima*)

From *Le Jardin du Roy Très Chrestien Loys XIII*, Paris, 1623

Engraving

Vallet's plants are presented with elegant formality, but are nevertheless botanically correct. The first edition of this book appeared in 1608 under Henri IV; it proved popular and was re-issued in 1623 with a revised title to take account of Louis XIII's accession.

presentation. Some publications fulfil both functions, being records of the contents of actual gardens, but with the illustrations intended as a source of floral motifs for designers. Pierre Vallet's *Le Jardin du Roy Très Chrestien Henri IV* (1608), as its title implies, depicts a selection of plants from the French royal gardens. The introductory text acknowledges the contributions of Jean Robin, Director of the Royal Gardens, who introduced a number of exotics from Spain and the islands off the Guinea coast. In terms of their naturalism and botanical content the engravings are equal to contemporary publications with a declared scientific purpose, such as Paul Reneaulme's *Specimen Historiae Plantarum* (Paris, 1611) and indeed far surpass a herbal such as Colombina's *Il Bomprovifaccia* of 1621 which re-used the long-outdated blocks from the *Ortus Sanitatis*. The 1623 edition attests to its botanical credentials by a dedicatory title-page incorporating portraits of the botanists Charles L'Écluse and Matthias L'Obel. Vallet worked in the gardens but also describes himself, in the preface to both the 1608 and 1623 (*Le Jardin du Roy Très Chrestien Loys XIII*) editions, as *Brodeur ordinaire* to the king. Furthermore the earlier book is dedicated to Henri's consort Marie de' Medici, who had a passion for flowers and set a fashion for embroidery with floral designs, so the book must be seen as primarily a source of motifs for the embroiderer. This is made explicit by Vallet who explains that the book was designed for those who paint (*peindre*), illuminate (*anluminer*), embroider (*broder*) and weave tapestries (*faire tapisserie*). As a further aid to the artist and designer using it as a source book the text is devoted to describing the colours of the flowers represented in these uncoloured plates. Similarly, a complete edition of the *Hortus Floridus* includes instructions for colouring the plates. Likewise Nicolas Robert's *Variae ac Multiformes Florum Species* (*c*.1660) was also intended as a source book for embroiderers, though Robert himself was a highly-regarded botanical artist who began the great series of watercolours on vellum (the *velins du roi*) that form the centrepiece of the collections in the library at the Jardin des Plantes. Another fine example of the botanically accurate pattern-book is the *Livre de Fleurs, Feuilles, et Oyzeaus* (1656), a collection of 27 engravings drawn from nature by Guillaume Toulouze, *Maistre Brodeur de Mont-Pelier*. Elegantly formal, these portraits of garden plants are nevertheless very finely detailed.

A compositional feature of the pattern-book type of florilegium is the bouquet, a usually-random mixture of flowers intertwined, or loosely tied with a ribbon. Nicolas Robert's early decorative work favours this device, and even an artist so firmly identified with the development of botanical science

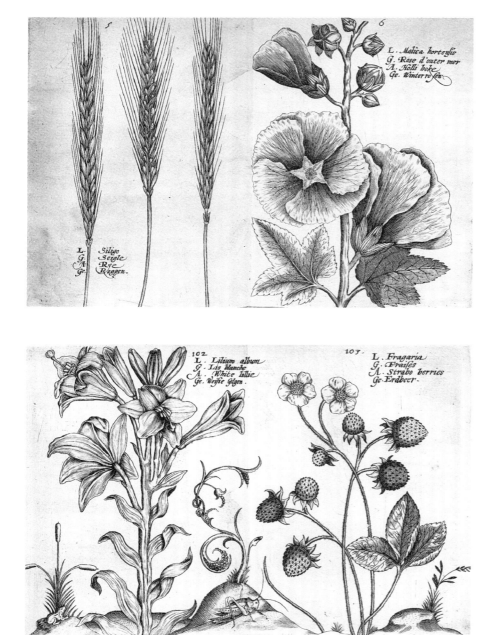

34 Crispin de Passe the Elder (1564-37)

Rye (*Secale cereale*) and hollyhock (*Alcea rosa*)

Plates 5/6 from *Hortus Floridus: Altera Pars*, 1614

Engravings

35 Crispin de Passe the Elder

Madonna lily (*Lilium candidum*) and strawberry (*Fragaria vesca*)

Plates 102/103 from *Hortus Floridus: Altera Pars*, 1614

Engravings

Many of the plant portraits in the *Altera Pars* (including the lily, hollyhock and strawberry shown here) are derived from extant watercolours by Jacques Le Moyne de Morgues, in the British Museum and the V&A. These formalized but botanically accurate studies are contrasted with elements of the purely decorative – the fanciful plant motifs or *rinceaux*, insects and animals.

36 Crispin de Passe the Younger (1597/8-post 1670)

Cyclamen sp.

Plate 13 from *Hortus Floridus*, Utrecht, 1614

The *Hortus Floridus* demonstrates the advance over woodcut that copper-engraving represented. The fine lines of an engraving can achieve more detail, and as here, can indicate tones and patterns by hatching and shading. This plate is richly tonal and scarcely needs colour to model the forms, though the introduction to the English edition gives details on how to colour the plates.

37 Crispin de Passe the Younger

Saffron crocus (*Crocus sativus*) and Crocus 'montanus'

Plate from *Hortus Floridus*, Utrecht, 1614

Engraving

The first edition of the *Hortus Floridus* had a text in Latin. It proved so popular that it was almost immediately followed by French, Dutch and English editions (the last was issued in Utrecht in 1615).

The plates are landscapes in miniature, embellished with animals and insects, and with the plants shown growing from the ground with a vigorous naturalism. The emphasis of the publication is on the common garden flowers, with a preponderance of spring bulbs.

38 Jean Louis Prevost (1760-1810)

Bouquet of paeonies, poppy, foxglove, pansies and French marigolds

Plate from *Collection des Fleurs et des Fruits*, Paris, 1805

Colour stipple engraving

This book was intended as a source of motifs for designers of china and textiles. Nevertheless the plants are naturalistically represented, though they are massed in arbitrary groups that disregard the flowering seasons of the component species.

as Georg Dionysus Ehret uses it throughout his career in many plant studies. In fact Ehret produced a publication which we might see as a late flowering of the florilegium form: the *Plantae et Papiliones Rariores* (1748-59). This is a strange hybrid production in which the plants are depicted with complete fidelity but presented in arbitrary fanciful bouquets, accompanied by an inaccurate text by Ehret himself. Pierre-Joseph Redouté also explored the decorative potential of botanical subjects in *Choix des Plus Belles Fleurs* (1827); each plate is titled '*Bouquet de…*'. Like Jean-Louis Prevost's *Collection des Fleurs et des Fruits* (1805) this was a sumptuous colour-printed equivalent to the floral pattern books of the seventeenth century, botanically accurate images with a primary decorative intent.

39 Georg Dionysius Ehret (1708-70)

Parrot tulip (*Tulipa gesneriana*)

Signed and dated, *G. D. Ehret pinxt. 1744*

Watercolour and bodycolour on vellum

This was one of the 44 illustrations Ehret contributed to C. J. Trew's *Hortus Nitidissimus* (1750-92), a publication devoted to garden plants. This was essentially a florilegium celebrating the beauties of flowering plants, and without any higher scientific purpose, as the subtitle asserts: 'The flower-garden in finest bloom throughout the year, or pictures of the most beautiful flowers.' This would explain the peculiarly exaggerated decorative character of the portrait, with its stylized rococo sinuosity.

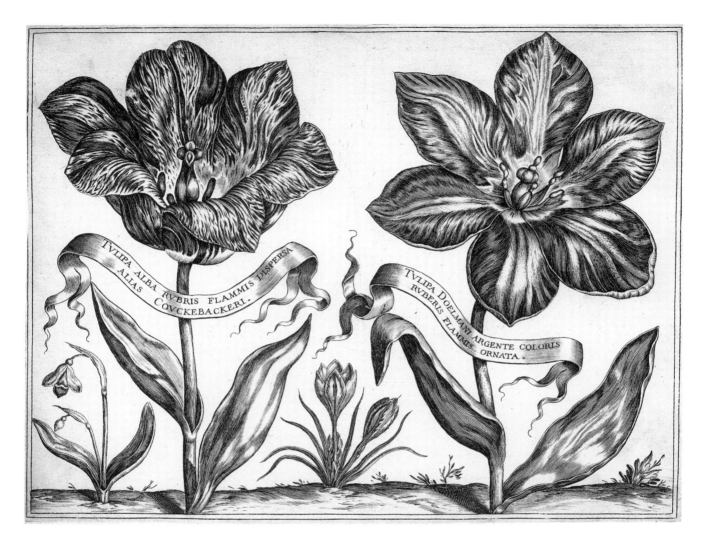

40 Langlois

Tulips

From *Livre de Fleurs*, Paris, 1620

Engraving

In the same fashion as Walther, or de Passe in the *Hortus Floridus*, Langlois presents his tulips somewhat truncated and growing in a flat landscape. Both flowers are tilted forward to give a clear internal view of the striped petals, the pistil and the stamens. Each plant is overlaid with a banner bearing its descriptive Latin name.

41 Guillaume Toulouze

Corn cockle (*Agrostemma githago*) and bearded iris (*Iris germanica*)

Plate 17 from *Livre de Fleurs, Feuilles, et Oyzeaus*, 1656

Engraving

This fine engraving comes from a book designed, as the title-page declares, to provide patterns for embroidery. It is nevertheless botanically exact, and detailed to the extent of delineating the stamens of the left-hand corn cockle flower, and the fine hairs on the buds.

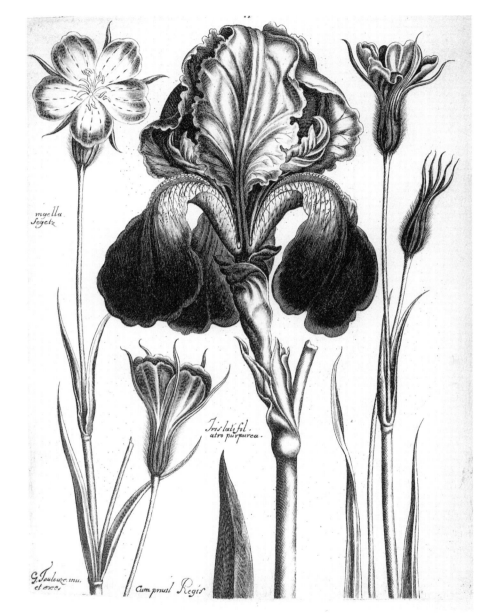

42 Georg Dionysus Ehret (1708-70)

Glade mallow (*Napaea dioica*) and morning glory (*Ipomoea* sp.)

Table VII from *Plantae et Papiliones Rariores* (1748)

Etching, coloured by hand

This odd hybrid publication, neither a proper botanical treatise (despite the inclusion of analyses of floral parts), nor precisely a florilegium, was a personal venture published, drawn and etched by Ehret himself. It illustrated exotics and new introductions in compositions which, though decorative, are eccentric and unscientific.

43 Pierre-Joseph Redouté (1759-1840)

Opium poppy (*Papaver somniferum*)

Plate 64 from *Choix des Plus Belles Fleurs*, Paris, 1827

Colour stipple engraving

Redouté learned about the potential of stipple engraving on his visit to England in 1786. He experimented with and perfected the method on his return to France. Following a successful court case in which he defended his rights to use it he declared:

'The process which we invented in 1796 for colour printing consists in the employment of these colours on a single plate by a method of our own. We have thereby succeeded in giving to our prints all the softness and brilliance of a water-colour, as can be seen in our Plantes Grasses, Liliacées, and other works.'
[Quoted in Blunt and Stearn, p.205]

This plate embodies all his claims for the process; it reproduces perfectly the fleshy sheen characteristic of the plant itself.

Pavot. *Papaver.*

P. J. Redouté. Langlois.

44 Tachibana Yasukuni (1717-92)

Poppies (*Papaver* sp.)

Plate from Part 3 of *Ehon Yazan-so* [Picture books of mountain and field plants], Osaka, 1806

Woodcut

Yasukuni's book has no pretensions to be a botanical treatise, though the plants are recognizable species, accurately rendered. The plants are labelled – in Japanese only – with their popular names. It seems to have been designed, as the title suggests, as a picture book for pleasure rather than instruction, the equivalent of a European florilegium. It was very popular, reprinted several times up to 1883 when colour blocks were added to the original line blocks from the first, 1754, edition. It is a fine example of the Japanese skill in using black and white alone to render form without lapsing into pure flat pattern.

New Discoveries

An important catalyst for the shift in botanical illustration away from the re-cycling of existing images to the production of original drawings from the life, or from preserved specimens, was the appearance in Europe of plants from Turkey and the Balkans, North Africa, the Americas, southern Africa, India and the Far East. Several writers struggled to identify these new plants with those described by the Classical authorities. Rembert Dodoens, in *Stirpium Historiae Pemptades Sex* (1583), in a typical attempt to assimilate the flora of the New World into that of the Old, identified tobacco with a European plant, Hyoscyamus, which also belonged to the family of Solanaceae and had been described by Dioscorides. Dodoens believed his two illustrated specimens to be images of the same plant, when in fact they were two distinct species later named by Linnaeus as *Nicotiana tabacum* and *N. tomentosa*. However, it soon became apparent that there were many more species in the world than could be encompassed by existing knowledge. With no authoritative images to draw upon, botanists commissioned drawings of these curiosities. There was also a new dependence on drawings because it became increasingly accepted that some plants would not grow in colder northern regions, and must be observed in their native habitat, and classified by means of dried specimens and drawings. New introductions demanded a higher standard of illustration than that established in most of the early herbals; those illustrations were adjuncts to a supposedly authoritative text, whereas an accurate description and classification of unfamiliar plants had to proceed from a faithful depiction.

The development of botanical science was profoundly affected by the discovery of new species, and the necessity of placing these unfamiliar plants in the existing scheme of things. Sir James Edward Smith, first President of the Linnean Society, described the problem that new discoveries thus posed for botanists:

> When a botanist first enters on the investigation of so remote a country as New Holland, he finds himself as it were in a new world. He can scarcely meet with any fixed points from whence to draw his analogies; and even those that appear most promising, are frequently in danger of misleading, instead of informing him. The whole tribe of plants, which at first sight seem familiar to his acquaintance, as occupying links in Nature's chain, on which he is accustomed to depend, prove, on a nearer examination, total strangers, with other

Helleborine Americana; radice tuberosa; folijs longis, angustis; caule nudo; floribus ex rubro pallide purpurascentibus.

Petro Collison Mercatori Londinensi.

E. Kirkall sc.

45 Helleborine (*Bletia purpurea*)

From John Martyn's *Historia Plantarum Rariorum*, 1728[-38]

Mezzotint printed in colours and finished by hand

This was the first botanical book illustrated with colour-printed plates, engraved in mezzotint by Elisha Kirkall after Jacob van Huysum and others. When skilfully done, as here, such plates required very little additional colouring. Each plate is dedicated to a patron-subscriber and has his coat-of-arms included. This incorporates the arms of Peter Collinson, an important collector of plants whose famous garden was much visited by botanists and artists, including Ehret. In fact Ehret produced a strikingly similar painted portrait of this plant in 1743 (Lord Derby's Collection), and must surely have been influenced by this earlier record of it.

Though this book pre-dates by several years the publication of the Linnaean system of classification, it nevertheless shows a significant interest in the floral parts, with three single flowers shown separately and from various angles, to give a fuller idea of the structure.

46 Maria Sibylla Merian

Passion flower (*Passiflora* sp.)

Plate 21 from *Over de Voortteeling en Wonderbaerlyke Veranderingen der Surinamsche Insecten*, Amsterdam, 1730

Engraving, coloured by hand

This work, first published in 1705, is in fact a treatise on entomology. Merian (1647-1717) spent two years in the Dutch colony in Surinam drawing from nature the insects and their food plants.

Though the plants are secondary to her real subject they dominate the plates and are of interest as some of the earliest representations of species from that region, and thus they have an important place in the history of botanical illustration. Nevertheless they are not botanically infallible: the copy in the National Art Library is annotated in a later hand with observations on Merian's botanical inaccuracies.

configurations, other oeconomy, and other qualities; not only the species themselves are new, but most of the genera, and even natural orders.[1]

It was this struggle to classify strange new plants that stimulated criticism, and ultimately an abandonment of the Linnaean system. Attempting to classify the flora he collected with Banks at Botany Bay, Dr. Solander filled his notebooks with erasures and revisions. Robert Brown, author of the definitive flora of the region, took the logical step of replacing the intractable Linnaean method with the natural system of de Jussieu; his decision led to a general acceptance of this system amongst British botanists.

It very early became established practice for artists to accompany voyages of exploration and colonization. The French painter Jacques Le Moyne de Morgues was the official artist with René Laudonnière's 1564 expedition to relieve the Huguenot settlement in Florida. Le Moyne mapped the sea coast and harbours, plotted the course of rivers, and had a brief to record anything worthy of note in the area. Given his existing and subsequent interest in flower subjects, it seems likely that he made drawings of plants on this trip, but most of his originals were lost in the chaotic circumstances of his escape from Florida.

The value of a painter on such voyages was widely acknowledged: in 1585 Richard Hakluyt the Elder tried to persuade the organizers of an expedition to North America that 'a skillful painter is also to be carried with you, which the Spaniards used commonly in all their discoveries to bring the descriptions of all beasts, birds, fishes, trees, townes etc'.[2] And the botanist Tournefort took Claude Aubriet with him on a botanical exploration of the Near East because 'without this help of Drawing, 'tis impossible any account thereof should be intelligible.'[3] In other words, pictures were essential to make sense of written accounts of scenes and subjects that were strange and new. And of course drawings had a further value because in the years of the early explorations it was immensely difficult to transport live plants: the voyages might last many months, and many plants died on these long sea-journeys, damaged by salt-spray and other uncongenial conditions. (In fact it was not until the invention of the Wardian case – a sealed 'Portable Greenhouse' for 'the conveyance of plants upon long voyages'[4] in the nineteenth century that there was a reliable means of transporting living plants safely). Although it was possible to bring back dried herbarium specimens, seeds, tubers and other organic materials, these things were more useful if supplemented by pictures of the mature plant in flower, together with an account of its natural habitat.

1 J. E. Smith, *A Specimen of the Botany of New Holland*, London, 1793, p.9.

2 Quoted in Ray Desmond, *The Wonders of Creation*, London, 1986, p.103.

3 Quoted in W. Blunt, *The Art of Botanical Illustration*, London, 1950, p.113.

4 *Companion to the Botanical Magazine*, Vol.I, 1836, p.319.

47 'Geranium Noctuolens Aethiopicum'
(*Pelargonium triste*)

Plate from Jacob Breyne, *Exoticarum Aliarumque Minus Cognitarum, Plantarum Centuria Prima*, Danzig, 1678

Engraving

This crowded and detailed plate is dominated by the dark bulk of the root, and framed by feathery leaves showing back and front views. With the flower-head half-obscured by the leaves it demonstrates how botanical illustration, pre-Linnaeus, placed no special emphasis on the flowering parts.

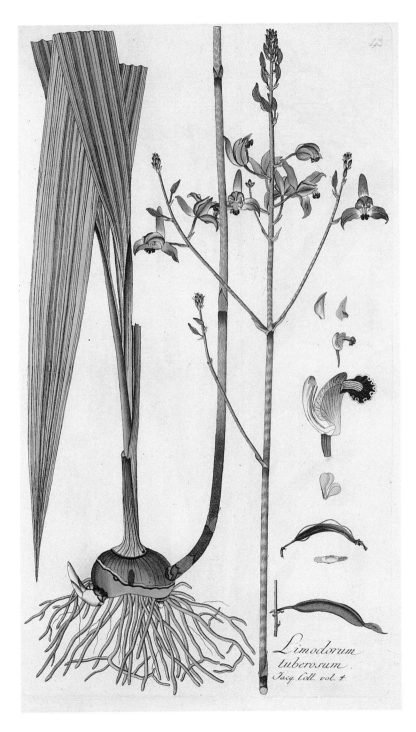

Limodorum
tuberosum.
Jacq. Coll. vol. 4.

48 Franz or Ferdinand Bauer

Limodorum tuberosum

Plate from Volume III of N. J. Jacquin, *Icones Plantarum Rariorum*, Vienna, 1781-93

Engraving, coloured by hand

This is one of the most impressive of Jacquin's many botanical works, and it is notable for being the only one with plates by Ferdinand and Franz Bauer. Jacquin is credited with having 'discovered' the Bauers and publishing their first botanical illustrations. Ferdinand went on to work on Sibthorp's *Flora Graeca* and Franz was persuaded by Sir Joseph Banks to work at Kew. The plates in this book, though unsigned, already show the qualities of detail, design and technical skill that characterize their mature work.

49 Walter Hood Fitch (1817-92)

Cyrtosia lindleyana

Plate XXII from J. D. Hooker's *Illustrations of Himalayan Plants,* 1855

Lithograph, coloured by hand

Botanically, Fitch's work was not as meticulous as that of Ehret or Redouté. In part this was a consequence of his phenomenal output – he is estimated to have produced more than 9,000 finished drawings and watercolours – but also the fact that he used lithography as his reproductive medium. This is well-suited to a swift broad style of drawing, but as the botanist John Lindley observed in his preface to *Illustrations of Orchidaceous Plants* (1830-38) 'even in the most skilful hands, is seldom adapted to high finish or delicate touch'. Nevertheless, Fitch's Himalayan illustrations were widely admired; the *Gardener's Chronicle* reviewer thought them the equal of drawings by the Bauers (though a comparison with pl. 57 will show that Fitch's drawing is much looser and more fluid).

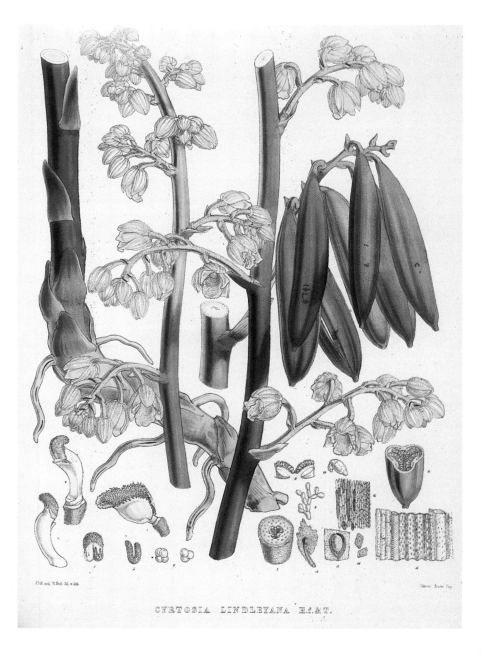

CYRTOSIA LINDLEYANA H.f.&T.

In some instances the artist was himself an amateur botanist. One of the earliest accounts of Japanese flora comes from Engelbert Kaempfer (1651-1716), physician with the Dutch East India Company. In 1790 when Kaempfer joined them, the Dutch (then the only foreigners permitted in the country) were confined to the island of Deshima off Nagasaki. Though Japan was a closed country and his expeditions were strictly limited to official delegations to the Court at Edo, he never missed an opportunity to gather 'plants, flowers and branches of trees, which I figured and described.' These drawings were published in his general account of his travels, *Amoenitates Exoticae* (1712), and later in the three-volume English translation of his *History of Japan* (1728). Plants newly-figured by Kaempfer included the maidenhair tree (*Ginkgo biloba*) which did not reach Europe until *c*.1730; and tea (*Camellia sinensis*), of which Kaempfer's drawing was the first European study at first hand. Kaempfer's illustration of tea (pl. 58) is particularly interesting because it includes his tiny copy of the representation of the plant in a Japanese herbal. Through this juxtaposition we are invited to read the Western view of the subject as pre-eminent, as being more accurate and of greater scientific value because it offers what purports to be objective analysis in place of subjective description.

A significant impulse to the discovery and introduction of new species from outside Europe was obviously an interest in their medical potential. Medicinal plants were the primary focus of the collecting activities of the botanical garden of the University of Amsterdam, for example. The director, Jan Commelin, and his nephew and successor Caspar Commelin, compiled a seminal illustrated catalogue – the *Horti Medici Amstelodamensis* (1697-1701) – of the many exotics introduced from the Cape, Ceylon, Brazil and elsewhere. Also important were new agricultural plants – maize, the potato, the tomato – all from the Americas; fruits such as the banana (Ehret in *Plantae Selectae*) and the pineapple. But the seventeenth century, which saw an unprecedented influx of new plants, was also the age of transition from potager ('vegetable plot') and physic garden to the pleasure garden, the garden as metaphorical recreation of Eden. Plants and gardens became status symbols, and exotic ornamental plants themselves became part of a much wider trade in luxurious commodities. The cargoes of the ships of the British East India Company and the Dutch equivalent returning from China, India and the Cape included plants as well as silks and calicoes, lacquer-work and porcelain. Private subscriptions supported lavish publications such as John Martyn's *Historia Plantarum Rariorum* (1728) which was devoted to new species growing at the Chelsea Physic Garden and the

OXYS BVLBOSA ÆTHIOPICA MINOR FOLIO CORDATO. *Fig. 22.*

50 'Oxys Bulbosa Aethiopica'
(*Oxalis incarnata*)

From Jan Commelin, *Horti Medici Amstelodamensis Rariorum Plantarum Descriptio et Icones*, Amsterdam, 1697-1701

Engraving

To give a true picture of this plant the artist has found it necessary to show it twice: first *en masse* to demonstrate its habit of growth, and second an isolated specimen to give a precise account of its structure.

Cambridge Botanic Garden. This is the first botanical publication with colour-printed plates; as in the *Catalogus Plantarum* which it pre-dates by two years, the plates were engraved in mezzotint by Elisha Kirkall after watercolours by Jacob van Huysum amongst others. Most of the eighteenth-century publications which present the rare, the new or the exotic were funded by wealthy individuals, themselves owners of important gardens. A good example is the *Hortus Cliffortianus* (1738), a catalogue of the plants in the Amsterdam garden of George Clifford, an Anglo-Dutch banker. The illustrations were by Ehret. As a consequence of family and professional links with the Chelsea garden Ehret was well-placed to produce many portraits of exotics and new arrivals. Around 100 of his finest studies were reproduced for Trew's *Plantae Selectae* (1750-73) including bananas, papaya, the night-flowering cereus, and the American Turk's-cap lily. Ehret's drawings are extraordinary for their lucidity and detail, but these plates produced over a period of nearly twenty years range in style from the naturalistic and pictorial to the abstract and diagrammatic.

Unlike the illustrations in herbals, say, or field guides, there is no unifying purpose or common style to distinguish illustrations of new discoveries. However there are some common characteristics and modes of practice in the representations of 'new' plants, whether they were produced in the seventeenth century or the nineteenth. We have seen how drawing from the living specimen changed the character of botanical illustration in the herbal for instance. This first-hand experience was often not available to the artist charged with drawing plants from the Americas and the East; at least not in the first phase of their discovery, before the plants were established in European gardens. Instead he was dependent on dried herbarium specimens or drawings by native artists from which he had to concoct a complete portrait of the subject. Jacob Breyne's *Exoticarum Aliarumque Minus Cognitarum Plantarum Centuria Prima* (1678) is an important and finely illustrated record of exotic plants from the Americas, the East Indies and South Africa. The engravings are by Isaac Saal after drawings by Andreas Stech and S. Cousins. The plate of *Leonitus leonorus* was based on a drawing sent to Hieronymus van Beverningk (to whom the book is dedicated) by Willem ten Rhyne from the Cape of Good Hope, where he stopped off en route to the East in 1673. Likewise the illustration of tea (*Camellia sinensis*) was derived from a drawing and seeds sent by ten Rhyne from Japan.

Many of the plants illustrated by Sir John Hill in his *Exotic Botany* (1759) came from China, and he gives an interesting account of his procedure:

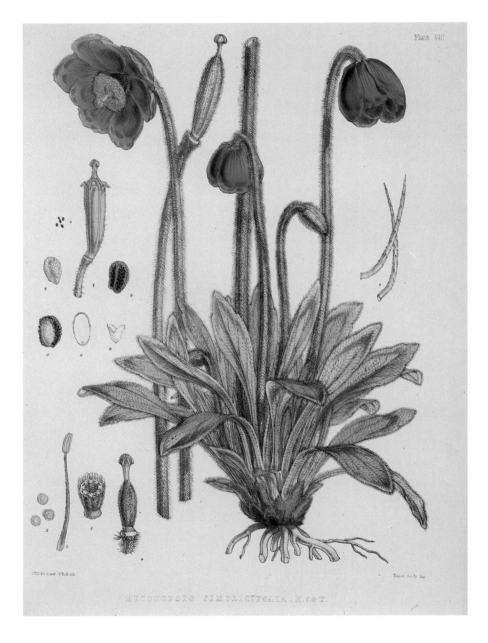

51 Walter Hood Fitch (1817-92)

Meconopsis simplicifolia

Plate VIII from J. D. Hooker,
***Illustrations of Himalayan Plants*, 1855**

Lithograph, coloured by hand

Hooker's *Himalayan Plants* illustrates those newly-discovered plants he thought worthy of cultivation in England. These dramatic and showy plants are well served by Fitch's confident and ostentatious style. He re-drew all the original drawings made by native Indian artists, correcting their perspectival faults and botanical inaccuracies in the process. This is an exception, one of the few plates drawn by Hooker himself, but also reworked and lithographed by Fitch. It is a comprehensive account of the plant, its structure and habit.

紅茉莉

52 Chinese artist (late eighteenth or early nineteenth century)

Plumbago indica

Watercolour

This watercolour, from an album of plant studies for the export market,
exhibits one of the characteristic features of Chinese botanical illustration: the
upper sides of the leaves are a solid dark green, the undersides a contrasting paler
tone. Some of the leaves are rather tired, curling and browning at the edges.
The influence of Western botany is manifest in the floral details, lower right.

Most of the plants came over dried, as specimens; and they were brought to the state wherein they are represented in these designs, by maceration in warm water. The method was this.

The plant was laid in a china dish, and water was poured upon it, nearly as much as the cavity would hold; another dish, somewhat smaller was turn'd down upon this, and the edges were cemented with common paste spread upon brown paper. This was set upon a pot half full of cold water, and placed over a gentle fire. Thus after a little time the lower dish heats; and the water gradually in it: a few minutes then complete the business. The plant, however rumpled up in drying expands and takes the natural form it had when fresh. Even the minutest parts appear distinctly.

The specimen is destroyed by this operation, but it shews itself, for the time, in full perfection: I could have wished to save some of these but they were sacrificed to the work; and I hope their remembrance will live in the designs.

Hill's methods are not dissimilar to those employed today: Stella Ross-Craig advocated boiling, or for very fragile material, gentle soaking. She also commented on the advantages and disadvantages of drawing from dried specimens: 'The artist gains…in not having to work at high pressure for a short period; but on the other hand there is the difficulty of creating the illusion of three dimensions. This can only be overcome by a thorough knowledge of botany and of perspective'.[5] Walter Hood Fitch (1817-92) however, a prolific and skilful illustrator who specialised in representing exotics, welcomed the challenge of drawing herbarium specimens. He declared that 'Sketching living plants is merely a species of copying, but dried specimens test the artist's ability to the uttermost; and by drawings made from them would I be judged as a correct draughtsman'.[6]

There is a tendency, evident in the illustrations to Breyne's book, for images based on herbarium specimens or made at second-hand from a drawing by an artist using non-European pictorial conventions, to be somewhat flat, and also tentative or overly specific; it is obviously difficult to create a generalized 'ideal' portrait from a single specimen.

Nevertheless an artist like Walter Hood Fitch could transform a dutiful and pedestrian study into a confident botanical portrait that convinces by virtue of its vigour and detail. Joseph Hooker's *Rhododendrons of the Sikkim-Himalaya* (1849-51) was illustrated with lithographs after the author's original slight field sketches, skilfully interpreted and translated onto the stone by Fitch. For Hooker's *Himalayan Plants* (1855) Fitch worked from

5 Blunt and Stearn, p.300.

6 *Gardener's Chronicle*, January-May, 1869.

LILIVM ZEYLANICVM SVPERBVM.

Fig. 35.

53 Glory lily (*Gloriosa superba*)

Plate 69 from Jan Commelin, *Horti Medici Amstelodamensis Rariorum Plantarum Descriptio et Icones*, Amsterdam, 1697-1701

Engraving

Commelin's book illustrates the plant introductions from the Dutch East and West Indies, South Africa and Ceylon, then growing in the Amsterdam Physic Garden. The plates are scientific in intent, and include botanical details, but there is a tendency to decorative embellishment, as in the spiralling flourishes to the tips of the leaves in this plate.

54 Chinese artist, late eighteenth or early nineteenth century

Unidentified plant of the leguminosae family

Watercolour and bodycolour

From the late eighteenth century European gardeners and plantsmen were increasingly interested in plants from China. Many of our most familiar garden plants such as tree paeonies, magnolias, and camellias were introduced from China in this period, as a consequence of trading contacts and plant hunting expeditions. Collectors and botanists commissioned paintings of Chinese plants from native artists in the trading ports of Canton and Macao. The artists were given examples of European illustration to copy and were trained in the western conventions of botanical drawing. They were adept copyists, but the Chinese drawings always retain a distinctive character that makes them unmistakable. The modelling of the forms is always rather solid, and done in two contrasting flat tones, as is evident from the stems and petals in this watercolour. The native Chinese style of painting involved abstraction and idealization of natural forms; when following Western instruction to give precise details in order that the plant might be identified and classified, the Chinese artist gave a literal transcription of the individual specimen with all its unique features. Thus in this picture each yellowing and withered leaf, each insect-bitten hole is precisely delineated, but expert botanical opinion suggests that the plant as a whole is as much a product of imagination as of observation.

55 Chinese artist (late eighteenth or early nineteenth century)

Peanut (*Arachis hypogaea*)

Watercolour

This drawing has the rather flat appearance which is so often a feature of the style of Chinese artists working in a Western idiom.

drawings by Indian artists commissioned by J. F. Cathcart of the Bengal Civil Service; Hooker stated in the Introduction that he used Fitch to correct 'the stiffness and want of botanical knowledge displayed by the native artists who executed most of the originals.'

This critical view of the skills of the native artists was not shared by all botanists engaged in the discovery and description of exotic flora. Robert Wright (1796-1872) produced four major floras of India and employed Indian artists in the process. He praised one of these men in particular for his 'skill in analytical delineation...as yet quite unrivalled among his countrymen and, but for his imperfect knowledge of perspective, rarely excelled by European artists'.[7] Many Indian artists were employed to produce drawings and watercolours for European botanists, plant collectors, and amateur natural-history enthusiasts. Their professional style was modified to accommodate European characteristics – they produced straightforward studies, the specimen isolated on a plain ground. There tends to be a certain formality of manner, the 'stiffness' of which Hooker complained.

Drawings of Chinese plants were also commissioned from native artists, mostly through the agency of the East India Company and its employees. Foreigners were confined to the island of Macao and allowed into Canton only when their ships were in port, so Europeans were necessarily dependent on Chinese artists. As early as 1698, James Cuninghame, a surgeon with the Company, had collected a variety of natural-history specimens, and acquired 'Paintings of near eight hundred plants in their Natural Colours, with their names to all, and Vertues to many of them'.[8] The East India Company itself also set out to acquire botanical drawings for its newly-founded London library and museum in 1804. The factory in Canton responded to the request:

> A botanical painter has been consistently employed in copying the plants, fruits and flowers of this Country, as they come successively in Season and we shall continue him till all that is curious in vegetable nature shall be designed. Mr. Ker,[9] His Majesty's Botanical Gardener, directs his employment and sends a description of those already painted which go in the Earl Camden's Packet together with Drawings of the Malacca Fruits by the same Artist.[10]

The Indian artists, working in a country where the British, as colonial power, had immense influence, were trained to satisfy the aesthetic and cultural standards of Western botanical art. The traditional methods and styles of Mughal miniatures were not to the taste of Europeans, and were ill-adapted to the demands of scientific botanical draughtsmanship. The solid

7 *Icones Plantarum Indiae Orientalis* vol.6, p.34

8 J. Petiver, *Museum Petiverianum*, 1699.

9 William Kerr, a Kew gardener, had been chosen by Sir Joseph Banks to be the first professional plant collector in China.

10 Letter of 29 January 1804, in the India Office Library.

gouache, the formality and symmetry, the lack of modelling and perspective, were replaced by Western practices: the use of flexible European paper, the exact copying of the plant in pencil or sepia ink, with subtly modulated washes of watercolour, using European illustrated books as models.

Unlike the Indian artists, the Chinese had to make few adjustments to their manner and techniques to satisfy the European botanist. Natural-history subjects were rare in Mughal art but there was a long tradition of naturalistic flower painting in China. Song dynasty (960-1279) painting in particular had been characterized by accurate detailed studies of birds and flowering plants. The Chinese artists were willing and able to supply the kinds of illustrations the plant collectors needed; their studies are characterized by an unfailing sense of design in the placement of the subject on the page, and a subtle, confident use of watercolour. Despite their ready assimilation of European demands, the Chinese drawings retain a distinct and unmistakable character. The technique for showing the undersides of leaves by using a paler shade of solid green was inherited from the native tradition. Another typical feature is the inclusion of blemishes, the marks of disease, age or insect damage, on the leaves or fruit. This is also held to be a traditional mannerism[11], but it may well be a consequence of the European insistence on fidelity and exactitude, drawing the specimen to hand and not idealizing the subject as would have been the natural procedure for a Chinese artist. These drawings were, after all, commissioned as records of new species to be sent back to Europe; it was essential that they be specific in order that the plant might be properly classified and named. Around 1800, both Indian and Chinese botanical studies begin to include separate details and dissections as the advances of European botany were belatedly communicated via prints and books. By imposing Western pictorial conventions such as perspective, the West colonized native perceptions of the flora, and effectively devalued 'other' ways of seeing and representing what was seen. The European style of illustration was eventually adopted everywhere, even in China and Japan where it supplanted the decorative though naturalistic imagery that had predominated in botanical books for centuries. It is in consequence of this that there is now a set of standard conventions, a universal graphic language for botanical illustration.

New discoveries were not always painted in their country of origin. Ehret and others painted new plants as they flowered in the botanic gardens and the collections of individuals: Ehret records drawings made in the gardens of Sir Charles Wager (the magnolia) and Peter Collinson (American

11 Craig Clunas, *Chinese Export Watercolours*, London (V&A), 1984, p.54.

EPIDENDRUM MACROCHILUM ROSEUM.

56 Samuel Holden
(*fl.* mid-nineteenth century)

Encyclia cardigera (*Epidendrum macrochilum* var. Roseum), 1844

Watercolour

Victorian plant collectors became fanatical about orchids; some enthusiasts amassed more than 18,000 plants. Holden painted many species in private collections: this drawing is annotated 'Mr. Farmer's, Nonsuch Park'. This orchid was first discovered on a plant-hunting expedition in Guatemala in 1837.

57 Franz Bauer (1758-1840)

Erica ramentacea

From *Delineations of Exotick Plants Cultivated in the Royal Garden at Kew*, London, 1796-1803

Engraving, coloured by hand

The subtlety and sophistication of Bauer's technique is exemplified in this detailed study. Heaths are notoriously difficult to illustrate because of their profusion of small flowers and tiny needle-like leaves, but Bauer has captured his subject with the finesse of a miniaturist. The plant is set out on the page with elegant clarity, its habit of growth and the stiff woody stems represented with simple, uncluttered accuracy. An illusion of three dimensions is created by Bauer's practice of using paler tones to indicate those leaves, flowers and stems which are overlapped by the foreground branches. The botanical perfection of the plate is completed with the magnified details of a full floral dissection.

Erica ramentacea

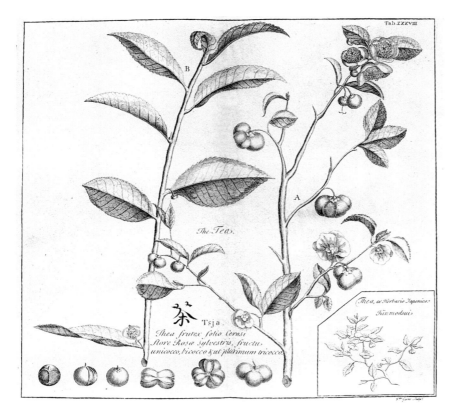

58 Englebert Kaempfer

Tea (*Camellia sinensis*)

From *The History of Japan*, translated into English by J. G. Scheuchzer, 1728

Engraving

Kaempfer's simple but competent drawings of the tea plant show the flowers and fruits (A) and a young shoot suitable for picking and drying to make tea (B). In contrast to his own sketches, which are consciously 'scientific' with their labelled parts and the magnified details of the fruits, Kaempfer has inset lower right a representation of the tea plant copied from a Japanese herbal. Allowing for the changes introduced in the process of copying, the Japanese image of the plant is nevertheless more decorative and sinuous than the Western 'objective' view.

Turk's-cap lily). At Kew, Franz Bauer painted his exemplary illustrations of the heaths introduced from the Cape region of South Africa, which are reproduced in his *Delineations of Exotick Plants Cultivated in the Royal Gardens at Kew* (1796-1803). Samuel Holden, who specialized in orchids, painted many of the species introduced from Asia, Mexico, Brazil and elsewhere, in the 1830s and '40s. His studies are often annotated with details of the collection in which he found his specimen; several were from the stock of Messrs. Loddiges of Hackney, nurserymen and publishers of the *Botanical Cabinet*. The practice continues to the present day with 'resident' artists at most major botanic gardens, including Kew and Edinburgh. It was Sir Joseph Banks who persuaded Bauer to stay in England as Kew's official botanical painter in 1790, where he remained for 50 years. Banks believed 'that the establishment of a botanic-garden cannot be complete, unless a resident draughtsman be constantly employed in making sketches and finished drawings of all new plants that perfect their flowers and fruits in it'.[12]

12 *Gentleman's Magazine*, 90, Pt.2, 381 (1820).

The Botanical Treatise:
Botanical Illustration and Plant Taxonomy

A preface to the era of classification can be seen in the seventeenth-century practice of juxtaposing, and thus ordering, natural specimens in collections, in herbaria and in gardens. In these spaces the specimens were ordered on the basis of their common features and surface resemblances. The collection is thus a catalogue or 'table' where the contents are ordered by visual criteria. With sight as the privileged sense for gathering knowledge of the world, then pictures of things take on a new value and a central role in the recording and communication of knowledge. The picture becomes an equivalent – a permanent, portable, intelligible equivalent – of the thing itself. Thus the botanical illustration became the medium of intellectual exchange in the development of botanical science; illustrations expressed theoretical positions quite as much as they described their ostensible subjects.

Until the seventeenth century, taxonomy, or the classification of plants, was haphazard and inconsistent, following a variety of conflicting theories. There was no agreed concept of species, no concept of varieties below the species level, no agreed terminology for the naming of plants, and indeed no agreement as to which physical features of a plant might be most useful or relevant in defining identity and relationship. One of the earliest attempts at using physical resemblance as a method of classification was made by Matthias L'Obel in his *Plantarum seu Stirpium Historia* (1576) in which he tried to arrange the plants in an order based on structural similarities; in some instances he anticipated the Linnaean system. Andreas Cesalpino (1519-1603), an Italian physician and philosopher, laid the foundations for all succeeding attempts to devise a 'natural system' of plant classification. In his *De Plantis* (1583) he recognized that

> Dioscorides' ordering [of plants] by medicinal properties, or the alphabetic arrangement used by others, are far from the nature of things. The best order is according to the community of natures. This is the best and easiest method for remembering, since characters proceeding from nature itself are recognisable and manifest to everyone, and are not deceptive like adventitious conditions which are impermanent. In this way an immense number of plants can be summarized in ordered classes.

In other words, a plant's stable physical attributes are the most apt and reliable features on which to base a system of classification; all other methods

HELLEBORUS niger, flore albo, etiam interdum valde rubente. ɪ.ɪ.
True Black Hellebore, or Christmas rose.

59 Georg Dionysus Ehret (1708-70)

Christmas rose (*Helleborus niger*) and winter aconite (*Eranthis hyemalis*), *c*.1745

Watercolour and bodycolour on vellum

Ehret shows the plants as growing, rather than as cut or uprooted specimens; this mode of presentation is anachronistic; it really belongs to the florilegia tradition (see Walther's watercolours, and the *Hortus Floridus*). The unseasonal butterfly also recalls the decorative bias of the florilegium. However, this naturalistic and decorative composition also incorporates everything then thought necessary for botanical analysis and identification: the stamens are clearly shown from several angles, in just-opened and mature flowers, and the flower itself is shown from the back, front and side.

60 Georg Dionysus Ehret (1708-70)

American Turk's-cap lily (*Lilium superbum*)

Watercolour and bodycolour on vellum

Painted in the 1740s, at the height of Ehret's mature style, this image exemplifies the heightened realism that he achieved in his finest work. Though Ehret had worked with Linnaeus since the late 1730s when they collaborated on the *Hortus Cliffortianus* (1738) the caption to this picture does not use the Linnaean binomial system but prefers the older, discredited method of using a string of descriptive Latin terms. However the Swedish botanist's influence is evident in the character of the illustration. The Linnaean system depended on the flower and an enumeration of the stamens and pistils, and their relation to one another. Thus Ehret has chosen to show only the flower-spike, ignoring the rest of the plant; the flowers themselves are painted in almost *trompe l'oeil* detail, from every angle, and in bud and full bloom. It is thus perfectly possible to classify this plant without ever seeing the thing itself.

In addition to the plant's descriptive name, Ehret includes a lengthy Latin description about the plant and the circumstances in which he studied it. He notes that 'this lily first flowered in August 1738, in the garden of Peter Collinson', an avid collector of new plants. It is not clear precisely when this finished portrait was made; this version is undated, but a near identical plate which appeared at Christie's in 1994 is signed, and dated 1745. Ehret often made copies of his more successful compositions, as commissions or gifts. One version of this was certainly sent to C. J. Trew, who published it in *Plantae Selectae* (1750-73).

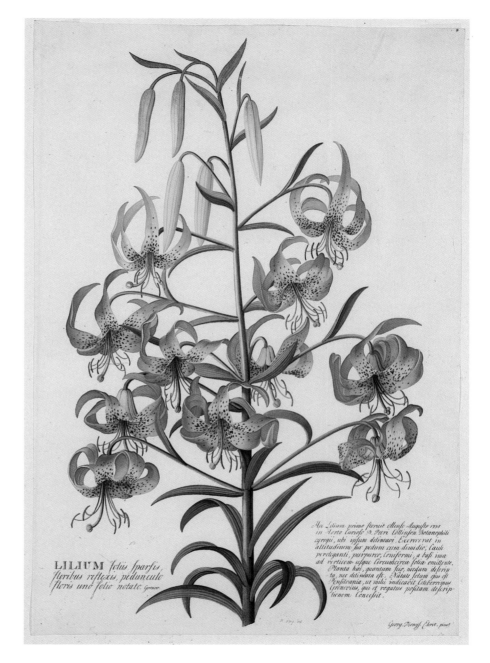

LILIUM *foliis sparsis, floribus reflexis, pedunculo floris uno folio notato.* Gronov.

are arbitrary, based on accidental and superficial features such as size, which might vary with soil, climate or habitat, or a plant's properties in relation to man, which are not essentially botanical.

Thus illustration, which had in the herbal tradition been simply a means of distinguishing one plant from another, now took on the role of analytic tool. Illustration, in the herbal, was secondary, descriptive; it could not represent affinities that were based on the intangibles of scent, taste, habitat or utility to man. But in the service of new theories it was used to record the detailed physical character of the plant and to express its morphological affinities. The herbal tradition had very largely depicted the whole plant without supplementary details but in the service of various taxonomists botanical illustration shifted its focus to accommodate and explicate differing theories.

There was a two-fold stimulus to the development of a systematic botany: the separation of the study of plants from medicine, and the influx of new plants which had somehow to be identified, described, named and illustrated. It was no longer possible to 'classify' or order plants by reference to their medicinal properties for these plants had as yet no place in a pharmacopoeia. And as we have already seen, it had already become apparent to Brasavola and others that there were many more plants in the world than could be identified with those described by the Classical authorities. It became increasingly evident that all plants must be fitted into a comprehensive scheme of classification, and so the selectivity of the herbal was replaced by an inclusive approach. In the herbal, only those plants which have an identifiable use are described and pictured; all others are excluded. (See Brunfels' apologetic inclusion of the pasque flower (pl. 10) and the lady's smock, and Gerard justifying the addition of the fritillary.) On the other hand, in the botanical treatise it is essential to include everything since all available evidence is needed both to construct and to prove the principles of a system of classification.

In the early eighteenth century the dominant system of classification was that of Joseph Pitton de Tournefort who recognized two types of genera: primary genera distinguished by their sexual organs, or 'fructification', and secondary genera, which shared a fructification pattern but differed in other characteristics. He divided flowering plants into 22 classes based primarily on the general form of the corolla. The theory was set out in the *Institutiones Rei Herbariae* (1700), illustrated by Claude Aubriet

61 Jan Wandelaar after G. D. Ehret

'Kiggelaria'

Plate from Carl Linnaeus, *Hortus Cliiffortianus*, Amsterdam, 1737

Engraving

This was Ehret's first collaboration with Linnaeus, and the excellence of his drawings certainly helped to promote Linnaeus's innovative botanical terminology and his theories of plant classification. Ehret however was somewhat offended by the lack of recognition for the value of his contribution, and especially his inclusion of floral dissections. He noted bitterly in his memoir 'Why did the others whose figures are in the same book not add the characters of plants? But I got no special honour for it and was treated as a common draughtsman'.

(1665-1742) whose drawings introduced the innovation of depicting details of seeds and reproductive organs separately. (Or more accurately, re-introduced, since both Gessner and Colonna were including details of flowers and floral parts in their figures in the sixteenth century). Tournefort's system became popular and was widely adopted because of its simplicity, and because of the excellence of the illustrations by which it was publicized. Increasingly illustration was fundamental to the development of botanical science; there could be no substitute for a picture in conveying quickly and unambiguously theories of classification that were founded on the physical character of plants.

By the mid-eighteenth century Tournefort's system had been superseded by that of Carl Linnaeus. In his *Philosophica Botanica* Linnaeus suggested that 'the essence of the flower rests in the anther and the stigma…[and the essence] of the plant in the fructification' so his classificatory method was based on the number of the sexual organs of the flower. Linnaeus's theories were publicized and promoted by means of illustration. The illustrator most closely associated with the dissemination of the Linnaean system was G. D. Ehret. Though Ehret contributed plates to the first Linnaean plant book, the *Hortus Cliffortianus* (1738), an account of the rare plants in the Amsterdam garden of banker George Clifford, Linnaeus complained that 'Ehret did in the beginning absolutely not want to paint the stamina, pistilla, and other small parts, as he argued they would spoil the drawing; in the end he gave in, however, and then he liked this kind of work so much that thereafter he observed the most minute and inessential particulars.' The book is one of the first to include floral dissections. It was also Ehret who drew the tabella or illustrative chart of the principle classes of the sexual system. This did much to popularize Linnaeus's ideas: it was reproduced by Linnaeus himself in editions of his *Systemae Naturae* and *Genera Plantarum*, and was widely pirated, appearing in several other botanical publications of the period.

Linnaeus's theories were adopted very slowly; they were not taken up by Philip Miller, for example, until 1759. Sir John Hill's *The Vegetable System* (1760-75) was one of the first British botanical works to use the Linnaean system, from Volume II onwards, and it survived in England long after it had been abandoned elsewhere. As late as 1831-4, a significant botanical book could appear under the title *A Selection of Hexandrian Plants* (by Mrs. Bury), 'hexandrian' in Linnaean terms being a plant with six stamens.

62 John Miller

Saffron crocus (*Crocus sativus*)

From *An Illustration of the Sexual System of Linnaeus*, [1770]-77

Engraving, coloured by hand

The eighteenth century saw a great proliferation of illustrated botanical books. The Linnaean system of classification was most influential and publications as diverse as his own *Hortus Cliffortianus* (1738) and Robert Thornton's *The Temple of Flora* (1799-1807) attempted to illustrate, explain and promote the Linnaean method. One of the most successful scientifically and aesthetically is Miller's *An Illustration of the Sexual System of Linnaeus*. This was published in various combinations of plates, coloured and uncoloured, between 1770 and 1777. The V&A has a number of proof plates before letters, of which this is one.

The advance on the herbal and the florilegium is obvious. Miller gives in a single plate a most complete account of the plant's habit, structure and life-cycle, with two views of the whole plant at different stages, together with dissections of the flower (vital to the Linnaean system), and details of seed-pod, leaf, corm and roots.

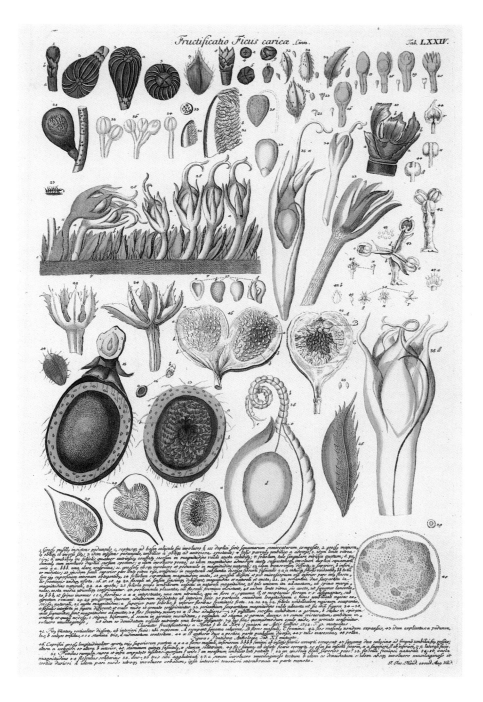

63 After G. D. Ehret

Fructification of the common fig (*Ficus carica*)

Table LXXIV from C. J. Trew, *Plantae Selectae*, Nuremberg, 1750-92

Engraving, coloured by hand

Unusually for Ehret, this is not a portrait of the whole plant supplemented with analyses (that appears in the preceding plate), but a page devoted to a full and detailed dissection of the reproductive parts – the fructification – of the fig. Some of the smaller parts have been left uncoloured, probably so as not to obscure the finer details. This plate exemplifies the then-current theories of classification which focused more or less exclusively on the structure of flower and fruit . The result is a composition of abstract beauty.

Robert Thornton's *A New Illustration of the Sexual System of Linnaeus* or the *Temple of Flora* (1799-1807) whilst purporting to be a serious botanical treatise was in fact closer to the florilegium tradition, and was of little value to the scientific enterprise of its time despite its impressive folio-size colour printed plates. The work is unconventional; the plants are not set against white or other neutral gounds, nor are they depicted in a convincing natural habitat. Instead they are set in elaborate Picturesque landscapes containing clues as to the country of origin. Thus the sacred Egyptian bean has the pyramids in the background; and the large-flowered sensitive plant (*Calliandra grandiflora*) is set in a Jamaican landscape 'with the humming birds of that country, and one of the aborigines struck with astonishment at the peculiarities of the plant'.[1]

1 Thornton, *Temple of Flora*, introduction.

A more scientifically-oriented work on the same theme, John Miller's *An Illustration of the Sexual System of Linnaeus*, appeared in parts from 1770 to 1777. The plates were somewhat extravagantly praised by Linnaeus himself as 'more beautiful and more accurate than any that had been seen since the world began'. His correspondent, naturalist John Ellis, wrote in a letter dated 28 December 1770 'the figures are well drawn, and very systematically dissected and described'.[2] The plates, arranged after the Linnaean classes and orders, are mostly of common plants, including the bramble and the sunflower. Full dissections of fruit and flower are included in most of the engraved hand-coloured plates, but again the focus is on the flower, with other aspects included arbitrarily.

2 See Smith, Sir J. E., *A Selection of the Correspondence of Linnaeus*, etc. 1, 255-6 (1821)].

The natural consequence of a classification system based on the sexual parts of the flower was a shift in the focus of illustration from the whole plant to the flower alone, or the flower and fruiting body. In the Linnaean system it was the characteristics of the flower which defined the genus; this was supplemented by other aspects of morphology, such as leaves, to determine the species. Perhaps the most notable feature of eighteenth-century Linnaean illustration is the reduction in the amount of information that it was felt necessary to include in a botanical drawing. Compare for example the illustrations from de Breyne, Reneaulme, Fuchs, even Besler, with some of Ehret's finest works. With the exception of a plate like the Christmas rose/winter aconite which unites the Linnaean pre-eminence of the flower with a representation of the growing plant in the tradition of the *Hortus Floridus* and other florilegia, Ehret's drawings focus almost exclusively on the flower. The American Turk's-cap lily (published in Trew's *Plantae Selectae*, 1750-73) is a case in point. The flowers are shown in almost

64 Sir John Hill

Marbled milk-thistle
(*Silybum marianum*)

**Plate 16 from *The Vegetable System*,
Volume 4, 1762**

Engraving

The 26 volumes of *The Vegetable System* (1759-75) comprise one of the most extensive British botanical publications of the eighteenth century. The first volume was pre-Linnaean but from volume 2 onwards Hill used the Linnaean binomials. The influence of Linnaeus is evident also in the inclusion of floral details to show the 'Generic character'. All the plates were 'Designed and Engraved by the Author'; he explains that 'Each figure represents such a portion of the Plant as contains all its characters: a flowering branch, with a larger leaf from the body of the stem, or root', all shown at a little less than 'what the parts would measure if laid upon the paper'.

For all his anxiety to show the characters and parts of the plants, Hill's sketchy graphic style tends to impose a uniformity on his subjects. With uncoloured plates such as this it is particularly difficult to distinguish between similar species.

65 Freret

Buphthalmum helianthoides

Plate XLV from C. L. L'Heritier de Brutelle, *Stirpes Novae*, Paris, 1784

Colour stipple engraving

A large proportion of the illustrations in this book are engraved after drawings by Redouté. Those which are the work of other artists are in his style; he also supervized the printing and colouring. It is a sophisticated composition, exploiting the rectangular page, and indeed emphasizing it with a border that is part of the original drawing, but not constrained by it. Certain details create a *trompe l'oeil* effect which enhances the illusion of volume. To the left one leaf is cut off by, and appears to pass behind, the 'frame', whilst another overlaps it. Likewise the details of the flower at bottom left and right extend out of the frame as if from a window. The way the shadows of the leaves fall across one another also contributes to the illusion that we are looking at a living plant, and not a flat diagram.

The plate is printed by colour stipple engraving, a technique promoted by Redouté and his 'school' because it was particularly effective in recreating the sheen on the leaves and petals in the original watercolours.

66 Johann Gesner

Class III. Triandria

Table 4 from *Tabulae Phytographicae*, 1795-1804

Engraving, coloured by hand

Gesner's work is an exposition of the Linnaean classes illustrated by tabular diagrams demonstrating the characteristics of each. 'Triandrian' plants are those with three stamens.

trompe l'oeil detail, each stamen and pistil distinct, and the flowers themselves ranging from buds to fully-mature blooms. It fulfils the Linnaean criteria perfectly, and has no need of a more comprehensive representation, so no indication of habit of growth, no root or bulb, no seeds.

The Linnaean focus on the flower produced much illustrative work that is closer to flower painting than to botanical art. Redouté is a good example; though he produced exemplary scientific work for botanists such as L'Héritier de Brutelle (a wealthy amateur who instructed him in the techniques of dissection and the details of plant anatomy) and Augustin Pyramus de Candolle, in which he gives all the details they demanded, he also worked independently of professional botanists. In his *Les Roses* and *Les Liliacées* he reverted to the Linnaean approach, often showing the flower only, and sometimes leaving out the dissections. Goethe, himself much occupied with botanical theory, felt that the demands of science were detrimental to the art in botanical illustration: writing in 1831 he declared that 'A great flower-painter is not now to be expected: we have attained too high a degree of scientific truth, and the botanist counts the stamens after the painter and has no eye for picturesque grouping and lighting'.[3] As he says, the 'great Low Country flower painter...had the easy task of pleasing the garden lover' whereas the botanical draughtsman knows that the botanical accuracy of his work will be examined by a whole host of critical experts. This division between 'flower painter' and 'botanical artist' has remained, though the same person may produce work in both categories. The annual exhibitions of the Society of Botanical Artists, for example, are dominated by works which, strictly speaking, are flower paintings – that is decorative though botanically-correct studies of plants painted for their own sake rather than for publication in a botanical journal or monograph.

3 *Conversations of Goethe with Eckermann*, 1984, p.327.

Linnaeus's other significant, and lasting, contribution to the development of botanical science was the establishment of a binary system of naming plants. This replaced the older method of naming which employed a string of descriptive terms in Latin, applied without a common standard or consensus. Many Linnaean names survive, but even where they have been superseded or there has been a reversion to the earlier name, the principle of allocating two names to a plant – the first to indicate genus, the second species, still stands. Linnaeus's contribution is commemorated in the authority citation given after the name; other authorities are likewise identified by an initial or an abbreviation.

Illustrations played an important role in the establishment of Linnaean nomenclature. The original specimen which is described to establish the species name and identity is known as a 'type'. Generally these are preserved in herbaria, but where the original type specimen is no longer extant, the drawing of it (or even a print, if that original drawing has not survived) may become the type. To differentiate drawings and prints from plant material, the former are often referred to as 'icontypes'. Linnaeus often based new species on published descriptions and figures alone and these illustrations are thus the original 'types' of a Linnaean species. Perhaps one of the best known examples is Brunfels' woodcut of the pasque flower (pl.10) Similarly it was the illustrations in *De Historia Stirpium* that established the new Latin-form names that Fuchs gave to Brunfels' *herbae nudae* – *digitalis* for foxglove, *ophioglossum* for adder's tongue and so on. This 'typification' of a botanical name remains an important function of botanical illustration.

Critics of Linnaean classification held that it was artificial and resulted in a system which grouped unrelated plants. Michel Adanson observed in his *Familles des Plantes* (1763-4) that 'The botanical classifications which only consider one part or a small number of parts of the plant are arbitrary, hypothetical, and abstract, and cannot be natural...the natural method in Botany can only be attained by consideration of the collection of all the plant structure'. Even Hill, who had initially so enthusiatically adopted the Linnaean system, became increasingly critical, and in the fifth volume of his *Vegetable System* proposed his own 'natural method' of plant classification.

A variety of theories of 'natural' classification were advanced: Antoine-Laurent de Jussieu, at the Jardin des Plantes in Paris, divided flowering plants into Monocotyledons and Dicotyledons, using the embryology of the plant as the basis for classifications, and largely ignoring the mature structure. Goethe's *Urpflanze*, or archetypal plant, represented the concept of metamorphosis, demonstrating how all plants' organs might be derived from a common structure. His ideas influenced de Candolle who modified de Jussieu's system with more morphological criteria – such as the relative positions of stamen and pistil.

John Lindley, following the theories of de Candolle, developed ideas on embryological characteristics such as seed structure as a means of classifying plants. He was the author of the *Botanical Register* whose plates, in the 1830s and 1840s, are generally superior to the *Botanical Magazine* because they included dissections and floral diagrams in accordance with the new

Large Flowering Sensitive Plant

67 After Philip Reinagle

'Large Flowering Sensitive Plant' (*Calliandra grandiflora*)

Colour aquatint for Robert Thornton, *Temple of Flora*, 1799-1807

Reinagle (1749-1833) was the most competent of the several artists employed by Thornton to paint the melodramatic oil originals from which his aquatinted plates were derived. Visually impressive, the illustrations are botanically deficient, for the artists were painters of portraits and landscapes, and none of them had any botanical training.

68 Pierre-Joseph Redouté (1759-1840)

**Bird of paradise flower
(*Strelitzia reginae*)**

**Plate from Volume 2 of *Les Liliacées*,
Paris, 1802-16**

**Colour stipple engraving with additional
hand-colouring**

Redouté's illustrations to *Les Liliacées* and
Les Roses (1817-24) constitute his best
known work. The title is misleading for its
eight volumes encompass lilies, irises, orchids
and other families.

Strelitzia Reginæ.   *Strelitzia de la Reine.*

theories. It was W. H. Fitch who introduced such analyses to the *Botanical Magazine*, from the late 1840s. Fitch was a consummate illustrator; his plates follow faithfully the descriptions appended, so for works with an ornamental intent he produced straightforward decorative representations, and for publications with a more scientific character the same plant would be accessorized with details and cross-sections.

Those illustrations which belong to specific theories of classification are apt to include only such data as are pertinent, and to exclude those aspects of the plant which play no part in that particular method of classification. Thus we find many illustrations which give only flower and fruit; roots for instance virtually disappear from botanical illustration after 1750, unless the plant is a bulb, or the roots themselves have some economic significance. The current practice is generally all-inclusive, as exemplified in the botanical work of artists such as Lilian Snelling, Margaret Stones, Mary Grierson, and Stella Ross-Craig. The latter's *Drawings of British Plants* (1948-73) show the whole plant, including roots, the flower and leaf, life-size where possible, plus sections and magnifications of all distinctive parts. The detailed analyses explain the structure quite clearly, and their consistency permits comparison between species and families. Ross-Craig refers [in the introduction to Volume I, 1948] obliquely to the long debate between the specific and the generalized in botanical illustration: not wanting to illustrate untypical specimens she examined examples in the Kew Herbarium to ensure that each plant she depicted was characteristic of its species.

Horticultural Illustration

On the evidence of the illustrations in horticultural publications, both scientific and commercial, colour is the most important feature of garden plants. Classification of garden plants is commonly based on colour variation, and it is the pre-eminence of colour which characterizes horticultural works. No purely horticultural publication survived on black and white illustrations: the most successful was Loudon's *Gardener's Magazine*, which was addressed to the working gardener rather than the amateur botanist. Founded in 1826, it failed in 1843, unable to compete with colour-plate rivals. Though botanical illustrations, on the other hand, are often coloured, they need not be, especially if their primary purpose is a description of plant morphology. Indeed unsophisticated colour printing or crudely-applied hand-colouring may obscure more than it elucidates. Sir Thomas Frankland, a subscriber to the *Flora Londinensis*, criticized the failures in the colouring of its plates as follows: '*Veronica officinalis* very good except the colouring of the flowers which is...done with opaque colours. I have not seen a single instance of opaque colours throughout your whole work without this spoiling the flowers – I had rather have no colour at all'.[1] Botany is concerned first and foremost with structure where horticulture – in the persons of the gardener, the florist, and the exhibitor – is concerned most immediately with colour. The obsessive cultivation of the tulip, and the other so-called 'florists' flowers' such as auriculas, anemones and carnations, and later pansies and dahlias, is bound up with efforts to produce new colours, or new combinations of colours (though there is a secondary interest in propagating double forms of flowers that more usually occur as singles). Throughout the *Floral Magazine* for instance, we find references to the colour of the plants represented: the Chinese primulas (pl. 80) are commended for 'such brilliancy of colour that they lay claim to being of the very first order' whilst of the poinsettia (pl. 82) it is said that 'The variety now figured is remarkable...for the novelty of its colour'.

Perhaps significantly, one of the earliest examples of colour printing is to be found in a de luxe nurserymen's catalogue. This is the *Catalogus Plantarum* (1730) produced by a 'Society of Gardeners', illustrating 'Trees, Shrubs, Plants and Flowers, both exotic and domestic, which are propagated for sale in the gardens near London...' The preface explains that the book will not include plants in public Botanic Gardens, or in the collections of

1 Curtis Museum, Alton, Hampshire. Curtis letters, item 116, February 16, 1781.

individuals, but 'only such as are actually in the Nurseries of the Persons belonging to this Society'. Only the first part, devoted to trees and shrubs, was actually published, but it includes several plates of flowers, intended for future volumes, which appear without text. Seven of the plates are mezzo-tints printed in colours – green and a reddish-brown – which were modified by additional hand-colouring. The printed colours are too even in tone, and rather 'soft', giving a misleading uniformity of texture to plants as various as the fleshy nasturtium and the waxy pine.

Until the end of the nineteenth century hand-colouring of outline-printed plates continued, even for serial publications, adding considerably to the expense. Obviously this kind of mass-production favoured styles of colouring that were simple, direct, and without superfluous and time-consuming subtleties of shading. Lithography was the favourite medium for these publications, among them the *Floral Magazine*, Jane Loudon's *The Ladies' Flower-Garden* series, and the *Botanical Magazine* from 1845 onwards. The lithographic process gives a soft and rather sketchy line not well-adapted for fine detail; in the style exemplified by the *Floral Magazine* the shading that indicates volume and depth is included in the drawing on the stone and thus printed. The colourist then simply has to add an even wash of colour rather than using graded tones of colour to modulate the forms. So as the illustrations themselves became increasingly formulaic, so too did the colouring. Lovell Reeve & Co., the publishers of the *Botanical Magazine* from 1845 to 1922, explained the procedures followed by their colourists in the 1920s: 'A regular colourer prefers to work in hundreds of the same plate, one colour at a time and one plate after another, in a purely mechanical way'.[2] The plant specimens illustrated are impossibly perfect. Just as the flowers are 'edited' to a perfection never found in nature, so their colours are painted in bold and lurid pigments which exceed their natural hue. This exaggeration of colour is even more pronounced in the printed horticultural material of this century. Seed packets from the 1930s and 1940s are almost luminous in their saturated colour. So too are the photographic illustrations that appear in the catalogues of bulb and seed companies, where colour is distorted by the use of lights, filters, and contrasting coloured backgrounds, and by the manipulation of the prints themselves to achieve an unnatural richness.

The earliest gardening books were not much interested in flowers, and the sparse illustrations are more often of garden plans and views, tools and implements, than of plants. Ferrari's *Cultura di Fiori* (1633; 1638) includes

2 Royal Botanic Gardens, Kew. Library (Archives). Publishers' notes on the production of Curtis's *Botanical Magazine*, January 11, 1922.

69 Christopher Switzer

Varieties of campion (*Lychnis* sp.)

From John Parkinson, *Paradisi in Sole Paradisus Terrestris*, 1629

Woodcut

The plants illustrated in Parkinson's book were mostly species growing in his own garden. It was designed as a practical book for the gardener rather than the scientific botanist. Though most of the illustrations are original they did not gain thereby; they are crude, stiff, schematic. The *Paradisus* is a relatively late instance of the use of woodcut for a botanical work given that it was not re-using existing blocks. By this date a number of significant publications (see pls. 3, 30, 34, 35) had established engraving as a superior medium for botanical illustration.

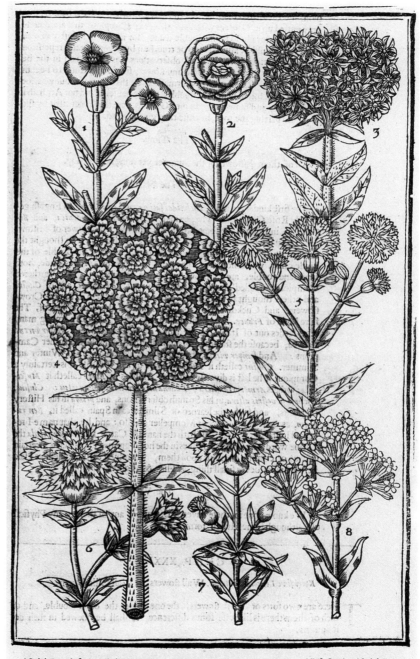

1 *Lychnis Coronaria simplex.* Single Rose Campion. 2 *Lychnis Coronaria rubra multiplex.* The double red Rose Campion. 3 *Lychnis Chalcedonica simplex* Single None such, or flower of Bristow. 4 *Lychnis Chalcedonica flore pleno.* Double Non: such, or flower of Bristow. 5 *Lychnis Plumaria multiplex* Pleasant in sight. 6 *Lychnis siluestris flore pleno rubro.* Red Batchelours Buttons. 7 *Lychnis siluestris flore pleno albo.* White Batchelours Buttons. 8 *Muscipula Lobelii.* Lobels Catch Flie.

a handful of simple images of bulbs and flowers, mostly lilies. De Cause's *De Koninglycke Hovenier* [The Royal Gardener] (1676) is illustrated in naturalistic pictorial fashion by richly-inked engraved plates in a landscape format, which show a strong influence (and in some instances direct copying) from the *Hortus Floridus*. As in the *Hortus Floridus* the artist employed a low viewpoint so that the flowers loom above the horizon and, seen in close-up, dominate the composition. This focus on the flower to the virtual exclusion of almost all other aspects of the plant becomes increasingly the defining feature of horticultural illustration.

John Parkinson's *Paradisi in Sole Paradisus Terrestris* (1629) follows the format of a herbal, listing and describing nearly 1000 plants, but unlike a herbal it gives instruction as to 'the right orderinge, planting and preserving of them' in addition to noting their uses and virtues. The illustrations to this 'choise garden of pleasant flowers' – woodcuts by Christopher Switzer – are original, not copied from elsewhere, and novel in composition: instead of following the established practice of the herbal tradition and depicting each plant as a single figure, or as in Besler's florilegium, setting out two or three to a page in elegant counterpoint, Parkinson's plates cram together a jumble of plants, many overlapping their neighbours. The effect is visually confusing though each plant is numbered to match against a key below. Parkinson claimed to have embellished the work with 'figures of all such plants as are materiall and different one from another: but not as some others have done, that is, a number of the figures of one sort of plant that have nothing to distinguish them but the colour, for that I hold to be superfluous and waste.' Yet to our eyes these crude uncoloured woodcuts, which have none of the subtle fineness of Fuchs, are often repetitive and barely distinguishable: the plates of daffodils and narcissi have little to differentiate the named varieties, and the distinctions between the tulips depend only on variations of hatching and spotting to indicate stripes since it is impossible to achieve gradations of tone in woodcut.

Many of Parkinson's plants – tulips, carnations, poppies – are shown as single flower-heads, most other plants as cut stems. Exceptions are the auriculas, cyclamens and irises for example where the whole plant is shown, some complete with roots or corms. The plant forms are much simplified, stiff and formal; they are often reduced to almost geometric regularity, or to formulaic structures (pl. 69) prefiguring the illustrations in many of the serial publications on horticulture that proliferated in the nineteenth century.

70 'Narcissus Indicus' (*Lilium* sp.)

From Ferrari, *Cultura di Fiori*, Rome, 1638

Engraving

A hybrid volume intended as a manual on garden design and the cultivation of ornamentals, *Cultura di Fiori* digresses into such subjects as how to make flowers bloom out of season and gives 'several ways of altering their form and colour'. It also gives early evidence of the use of flowers indoors, with advice on the most appropriate occasions for flowers, and how to compose bouquets. The illustrations are a mixture of the decorative and scientific, embellished with baroque elements such as festoons and ribbons inscribed with the plant's name. Yet it also contains a very detailed description of the 'Chinese rose' or hibiscus, first grown in Rome by Ferrari, accompanied by an illustration of the seed as seen through a microscope.

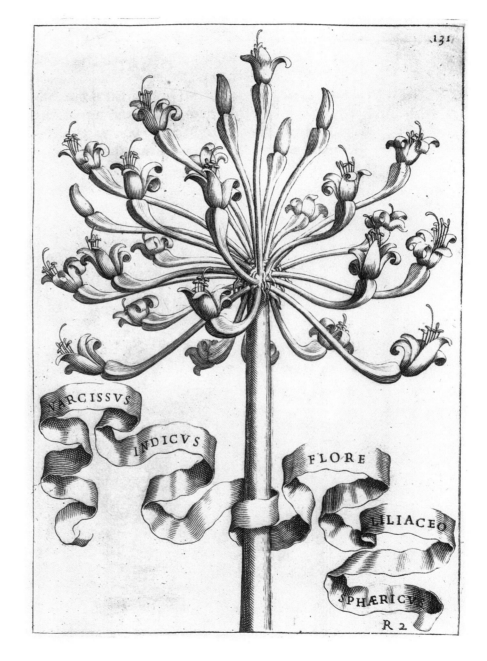

The common feature in horticultural illustration is the emphasis on the flower to the exclusion of almost everything else. The details and dissections essential to botany are superfluous to horticulture. The sexual parts of the flower are often indicated in the most cursory fashion that would be completely inadequate for the botanist (pl. 80). The horticultural illustrator presents a dominant flower and presents it frontally. This flattening process and frontal orientation can be seen in the series of prints after flower pieces by Peter Casteels which the Kensington nurseryman Robert Furber used as a month-by-month catalogue of flowering plants available from his stock. Here the natural form of a bouquet, its volume, is denied, and every flower faces front, arranged separately and distinctly from its companions. The prints were re-issued many times, on a smaller scale by different London book- and print-sellers, often as purely decorative plates but sometimes with genuine horticultural purpose. In 1732 they appeared in an anonymous quarto volume, engraved by James Smith, as *The flower-garden display'd, in...four hundred...representations of the most beautiful flowers; regularly dispos'd in the respective months of their blossom...engraved on copper-plates from the designs of Mr. Furber and others... With the description and history of each plant.* The author admitted his publication was motivated by the fact that 'the first impression of the monthly flower-pieces have been so well-received by the publick' but they are inferior copies of Furber's originals and have little artistic merit. However, with the addition of bold if crude hand-colouring they seem to have answered the needs of the plant-buying public. In this version they are supplemented with a text giving 'a particular description of the flowers, and the nature of their culture, which could not be expressed in the plates themselves', and are thus decorative additions to a gardening manual rather than illustrations to a trade catalogue.

The frontality initiated in the Furber prints appears in ever more exaggerated forms in a whole succession of nineteenth-century horticultural publications such as *The Florist*, the *Floral Magazine*, the *Botanists' Repository*, and the *Botanical Register*. There are three styles of illustration which characterize these floral magazines and catalogues: the specimen portrait in which the artist shows an individual flower, the stem and at least one leaf; the composite portrait – a bunch or bouquet of flowers, showing variant colours or forms; and the diagrammatic – a single flattened geometric form, either in outline or coloured.[3] The specimen portrait is found in Benjamin Maund's *The Botanic Garden* (1825-51); in some instances where the fruit is also decorative it is included but there are no other botanical details. Each

3 I am indebted to Brent Elliott, of the RHS Lindley Library for the definition of these categories given in his paper to the RHS Conference on Botanical Illustration, Edinburgh, July 1994.

small plate (there are four to a page) does however incorporate some useful information for the gardener: some images are accompanied by a monochrome sketch of the whole plant to denote those which are shrubby. Every plate has an indication of scale such as ½ or ¼ life-size given in the lower right corner.

The composite style is demonstrated in many plates in the *Floral Magazine*, such as the 'Single Varieties of Chinese Primula', the 'Fancy Pansies', or 'Varieties of Verbena'. Some of the most attractive examples appear in Jane Loudon's *The Ladies' Flower-Garden Ornamental Annuals* (1842) where, rather than grouping together hybrids or colour variants of the same plant, she creates an informal bouquet of plants from the same family, such as the convolvuluses and morning glories (pl. 78).

The diagrammatic manner is the most stylized, with the flower reduced to a circle, internally subdivided to indicate separate petals and disposition of colours. All three styles are subject to reductive mannerisms and abstractions that smooth out and perfect. The artists who illustrated these publications worked to formulae; just as the gardener aimed to improve on nature, breeding bigger, better, brighter flowers of regular shape and clearly defined colour combinations, so the illustrator produced round, regular, crisp, flawless, evenly-coloured representations.

Less prescriptive and stylized but equally concerned with well-coloured illustrations of showy exotics was Curtis's *Botanical Magazine*. This was perhaps the most successful (and certainly the longest lived) of the horticultural serials. First published in 1787 and still surviving as *The Kew Magazine*, it was to show 'the most Ornamental foreign plants cultivated in the Open Ground, the Green-House and Stove'. These were illustrated in new figures drawn always 'from the living plant and coloured as near to nature, as the imperfection of colouring will admit.' Curtis in his introduction describes it as 'a work in which Botany and Gardening...might happily be combined'. For a long time it kept its predominantly horticultural character but it became more scholarly, botanical details were added to the plates and in J. C. Loudon's words 'elegant science' replaced 'floral amusement'. The earliest volumes are modestly illustrated by hand-coloured engraved plates drawn by Sydenham Teast Edwards (1768-1819) in a style that is closer to the delicacy and detail of Curtis's other important work, the *Flora Londinensis* (for which Edwards also drew), than it is to the mostly-schematic, highly-coloured imagery typical of the horticultural series of the

1860s and 1870s. (The plates were engraved on copper until the end of Volume 70; thereafter lithography was employed until the introduction of colour printing in 1948.) However, by the 1840s, under the editorship of W. J. Hooker, the plates were exclusively the work of the prolific Walter Hood Fitch. The focus of the periodical had shifted from the ornamental garden plant to the rare, showy and often delicate exotics grown at Kew (*Alloplectus capitatus*, pl. 75) and in specialist nurseries, or shown at the Horticultural Society (*Medinilla magnifica*, pl. 76). Fitch introduces botanical details to many, though not all, of his lithographed hand-coloured plates. Fitch's style is fluid and confident, with a strong sense of design, and the drawings are perfectly complemented by the work of the colourists.

Another periodical noted for the high quality of its printed illustrations was the *Transactions of the Horticultural Society* of London (1807-48) which employed one of the many skilled female illustrators, a Miss Drake (of whom surprisingly little is known although she worked on major publications for Bateman and Lindley). Although there had been noted female flower painters (Mary Moser, Rachel Ruysch) and illustrators (Madeleine Basseporte, Maria Merian) it was only in the nineteenth century that significant numbers of women were professionally employed on botanical illustration. Many of the serial publications were dependent on female artists: Maund's *Botanic Garden* included Mrs. E. Bury, Miss E. Maund and Miss S. Maund amongst its contributors, and Loddiges' *Botanical Cabinet* cites among others Miss J. Loddiges and Miss Rebello. In the twentieth century the field of botanical illustration has been dominated by women, from Lilian Snelling and Stella Ross-Craig to Margaret Stones, Mary Grierson, Ann Farrer, Pandora Sellars (whose work was reproduced on a set of Royal Mail stamps issued in March 1993) and Jenny Brasier. Brasier, who has perfected the now-rare practice of painting on vellum, has done a good deal of horticultural work, with contributions to Diana Grenfell's monograph on hostas, *The RHS Dictionary of Gardening*, 1992, and an ongoing series of plates for the *Cyclamen Society Journal*.

Illustrations have played a significant part in publicizing and popularizing new introductions and hybrids. The horticultural periodicals both reflected and fostered the changing fashions in garden plants. Nowadays this role is fulfilled largely by the catalogues of nurseries which draw attention to new varieties and new colours – the 'blue' rose, the red delphinium – and tempt customers to buy with vivid pictures of perfect blooms. Photography is now generally preferred both for gardening books and for commercial

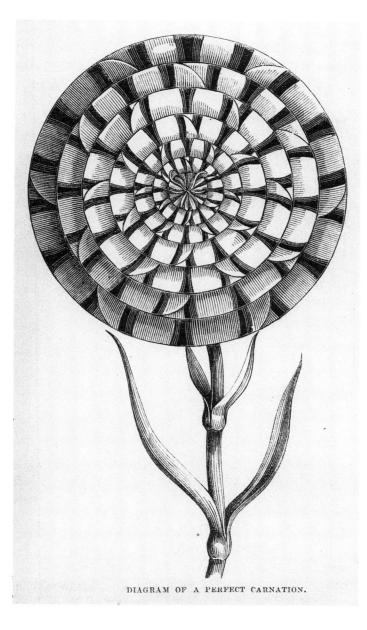

DIAGRAM OF A PERFECT CARNATION.

71 Diagram of a perfect carnation

From the *Gardener's Magazine of Botany*, July-December 1850

Wood engraving

This is a precise geometric diagram of the flower intended to demonstrate the perfection of form to be aimed for by gardeners. As the text puts it, the ideal is a 'circular roseate flower, the more round the outline the better.'

horticultural literature; photographs have a lush, seductive quality, and an immediacy perhaps, where a drawing suggests a more scientific approach and is thus more usually the preferred medium of the botanical monograph and field guide.

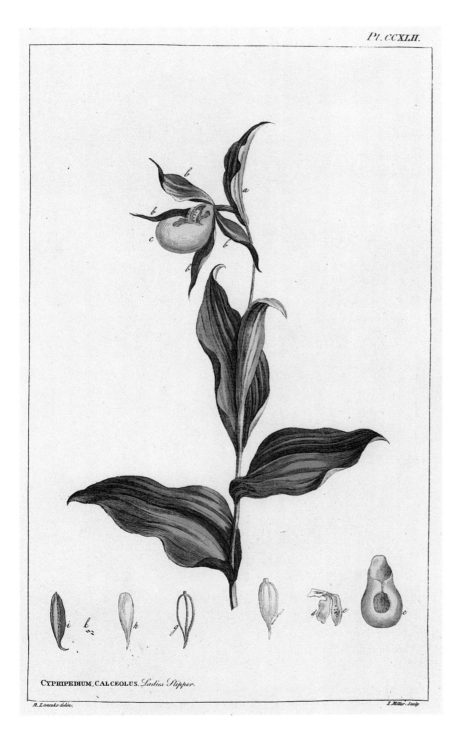

72 After Richard Lancake

Yellow lady's slipper orchid (*Cypripedium calceolus* var. *pubescens*)

Plate CCXLII from Philip Miller, *Figures of the Most Beautiful, Useful and Uncommon Plants in the Gardener's Dictionary*, third edition, London, 1809

Engraving, coloured by hand

Miller's original *Gardener's Dictionary* was first published, without illustrations, in 1731, followed by the first illustrated edition in 1755-6. It was a popular publication with handsome plates after Ehret, John Miller, and others. It illustrated chiefly 'those Plants which are either highly ornamental and remarkable for the singularity of their structure, or the beauty of their flowers', though it also included some of primarily medicinal value. The plants are classified by the Linnaean system with 'Generic characters…fully and accurately delineated'.

73 After Peter Casteels (1684-1749)

August from *The Twelve Months of Flowers* (1730-32)

Engraving, coloured by hand

This is one of a set of engravings representing the months, each in the form of a bouquet of flowers in an ornamental vase. Peter Casteels, a Flemish painter of flower pieces, produced the original painted versions for Robert Furber, a nurseryman and gardener of Kensington, London. Furber had them engraved and published as *The Twelve Months of Flowers* (1730), and they served as a catalogue of the plants for sale from his nursery. This was the first seed catalogue produced in England. Each flower is labelled with a number, and there is a key below giving the names of the plants. They are ordered approximately by season of flowering. The large scale versions are usually hand-coloured, but the compositions became popular as decorative prints and were reprinted first as *The Flower Garden Displayed* in 1732, and a further six times between 1740 and 1760 in smaller uncoloured editions.

The decorative potential of the images was recognized in the frontispiece to *The Flower Garden Displayed*, which describes them as 'Very Useful, not only for the Curious in Gardening, but the Prints likewise for Painters, Carvers, Japaners, etc., also for the Ladies, as Patterns for working, and Painting in Water-Colours; or Furniture for the Closet'.

The prints are composed following exactly the conventions of the Dutch flower piece, with the flowers arranged frontally, each distinct and separate. Natural growth and length of stem are subordinated to the demands of a balanced arrangement. The painted model is imitated to the extent of including a scattering of loose flowers on the ledge which supports the vase.

74 After Jacob van Huysum

Double nasturtium (*Tropaeolum majus* 'Flore pleno'), Spanish tree germander (*Teucrium latifolium*) and dwarf colutea (*Coronilla glauca*)

Plate 3 from *Catalogus Plantarum*, London, 1730

Colour mezzotint

This book, with its mezzotints by Kirkall after Jacob van Huysum (*c.*1687-1740) (see pl. 45) was a prestige catalogue produced for the twenty named members of a Society of Gardeners. All were nurserymen, and the plants illustrated were rare decorative garden varieties from their stock.

In terms of composition this, and other plates in the *Catalogus*, are strikingly similar to the florilegia of a century earlier. The format in which a showy central plant is accessorized by smaller and more modest specimens is first seen in Besler's *Hortus Eystettensis* of 1613 (pl. 31).

75 Walter Hood Fitch

Alloplectus capitatus (1817-92)

Table 4452 from Curtis's *Botanical Magazine*, Volume V, Third Series, 1849

Lithograph coloured by hand

This boldly-coloured plate demonstrates Fitch's compositional skill – he utilizes every corner of the rectangle, taking the image right to the edges, yet the plant itself is not rectangularized by the confines of the plate.

Fitch del et lith.

F. Reeve, imp.

2.

76 Walter Hood Fitch (1817-92)

Medinilla magnifica

Table 4533 from Curtis's *Botanical Magazine*, Volume VI, Third Series, 1850

Lithograph, coloured by hand

Fitch contributed his first plates to the *Botanical Magazine* in 1834, and from the 1840s to 1878 was its sole illustrator. The fact that he also lithographed his own work ensured that his drawings lost none of their vitality in the printing process. This plate is only partly coloured, with all but a small portion of the leaves left in outline, so that their dark solid green does not overwhelm the delicate form and colour of the flowers. Fitch in fact makes a virtue of necessity: limiting the colouring of the plates to the inflorescence and a single leaf was an economy imposed by the publishers. To have the part stand for the whole is a common device in botanical illustration when working in a restricted space, because it ensures that the composition remains uncluttered and readable.

This plant is represented on a folded plate, since the regular format could not properly accommodate it. Curtis had introduced such plates with reluctance, regarding them as a deviation to be indulged in only for something 'uncommonly beautiful or interesting' such as the *Magnolia liliifolia* or the *Strelitzia reginae*, but they became more common under Hooker's editorship.

77 After Miss S. A. Drake (*fl.*1830s-40s)

Cattleya guttata

Plate 8 from the Horticultural Society's *Transactions*, Volume II, Second Series

Aquatint, coloured by hand

Miss Drake seems to have specialized in orchids. She contributed many of the plates to Bateman's monumental *Orchidaceae of Mexico and Guatemala* (1837-41).

1. *Convolvulus elongatus.* — 2. *Convolvulus tricolor.* — 3. *Ipomæa bona-nox.* — 4. *Convolvulus siculus.*
5. *Ipomæa barbigera.* 6. *Convolvulus purpureus var. elatior.* 7. *Ipomæa rubro-cœrulea.* — 8. *Convolvulus involucratus.*
9. *Ipomæa coccinea.* — 10. *Ipomæa Quamloclit.*

78 Jane Loudon (1807-58)

Convolvulus and morning glory species

Plate 26 from *The Ladies' Flower Garden of Ornamental Annuals*, London, 1840

Lithograph, coloured by hand

Jane Loudon's book was one of a series she devoted to garden plants. The importance of colour for differentiation is obvious in a plate such as this, which groups plants that are very similar in form. Until the advent of cheap colour printing, hand-colouring remained essential to the success of horticultural publications.

79 James Andrews

Dahlia Mrs. Bush
(*Dahlia superflua*, var.)

Plate 88 from the *Floral Magazine*, Volume II, 1862

Lithograph, coloured by hand

The ideal dahlia is presented here as perfectly round, its petals tight and crisp. Andrews has sketched it with an almost geometric evenness and simplicity, and the bold globe of the flower dominates the composition. The text makes clear that the image represents the 'progress' of the dahlia to a greater perfection of form and contour, and also cites the colour (which should be a 'soft peach', not the shocking pink the colourist has used) as an important feature. It is apparent in the leaves, stems and bud, especially, how the forms are modulated by the artist drawing directly on the lithographic stone, so that the colourist has only to add even washes of pigment to complete the picture. The colouring is itself limited to three shades of green, one each for the upper surface of the leaf, the underside, and the stems.

J Andrews, del. et lith.

Vincent Brooks, Imp

SINGLE VARIETIES OF CHINESE PRIMULA.

FLORAL MAGAZINE NEW SERIES.

L. Reeve & Co. 5. Henrietta. St. Covent Garden.

80 Single varieties of Chinese primula (*Primula sinensis*)

No.18 from the *Floral Magazine*, 1872

Lithograph, coloured by hand

This is a good example of a frontal composite bouquet, showing a number of variant colourways. Each is numbered with a key in the accompanying text. The forms of the plants are very much simplified; often structure is merely suggested rather than being precisely delineated, and the addition of colour tends to obscure the finer detail of the drawing. It is this broad 'impressionistic' approach and exaggerated colouring which distinguishes much horticultural illustration from the more strictly botanical.

81 Fancy Pansies (*Viola* sp.)

No.29 from the *Floral Magazine*, 1872

Lithograph, coloured by hand

This demonstrates how formulaic much horticultural illustration was in the nineteenth-century serial publications. The flowers are reduced to simple flat geometric forms defined by solid blocks of colour.

The essentially decorative quality of the *Floral Magazine's* illustrations was evident after it ceased publication in 1881: for some years Lovell Reeve, the publisher, sold off the stock of loose plates for screens, scrapbooks, studies in flower-painting etc.

W.G.Smith,F.L.S.del et lith.   V.Brooks Day & Son,Imp

FANCY PANSIES.

1. "James Neilson."   2. "Lady Ross."
3. "John B. Downie."   4. "David Mitchell."

POINSETTIA PULCHERRIMA. rosea-carminata.

W.G.Smith.F.L.S.del.et lith.   V.Brooks,Day&Son,Imp.

82 Poinsettia (*Euphorbia pulcherrima*)

Plate from the *Floral Magazine*, 1873

Lithograph, coloured by hand

The *Floral Magazine* was published by Lovell Reeve, who also owned the *Botanical Magazine*; in contrast to the latter it was devoted to 'such introduced plants only as are of popular character, and likely to become established favourites.' Reeve stressed that this was to be reflected in the illustrations. Writing to Thomas Moore, the editor, in 1860, he explained that 'Having the Botanical Magazine we not only do not want botanical subjects, but we want floral subjects treated both by editor and artist more *florally*.'* Moore, and Fitch (the artist for the first sixteen issues) were asked to resign and were replaced by the Rev. H. H. D'Ombrain as editor and James Andrews as artist; they simplified both text and images, giving the publication a more decorative and popular character.

The strong design of this plate focuses on the most colourful part of the plant, the tiny, almost insignificant flowers and the bright red bracts. Colour is emphasized as the *raison d'etre* for cultivating the plant, and for representing it.

* Royal Botanic Gardens, Kew Library (Archives). Lovell Reeve. Loose letters, item 323, May 11, 1860.

83 (Clockwise from upper left)
Kalosanthes coccinea, Pavia rubra,
Daphne mezereum and *Cotoneaster*
rotundifolia.

Plate 305 from Volume XIII of Benjamin
Maund's *The Botanic Garden,* early
1830s

Engraving, coloured by hand

This shows the standard format for the plates
in the serial publication *The Botanic Garden*
with four species to a page. Maund's work
was essentially a catalogue of garden plants,
so the focus is on the flowers, with the
addition in the case of the cotoneaster of the
decorative red berries. A tiny monochrome
sketch shows that the daphne is a tree-like
shrub. The delicate, pale flowers of the pavia
are set against an outline drawing of the leaf
which gives a useful indication of relative size
and scale. Each of these less-than-life-size
pictures has an indication of the scale given
lower right.

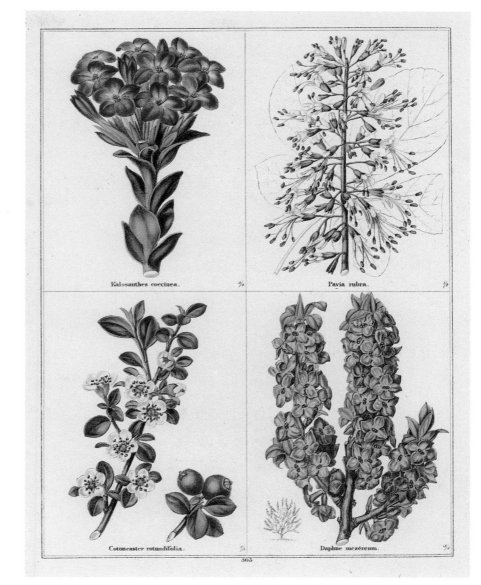

84 A. Sunzer

Poster advertising Mandeville & King Co., of Rochester, New York, suppliers of flower seed, 1907

Colour lithograph

The bouquet motif, so often employed in floral pattern books (see pls. 39, 42) was also favoured for horticultural illustration. It emphasizes the decorative role of garden flowers, both in the garden setting (two subsidiary vignettes show the plants growing) and in the domestic interior. Here the bouquet is informal, with the simple charm of a child's bunch of flowers, suggestive more of natural abundance than of manipulated perfection, and in that sense unusual in horticultural illustration which tends to exaggerate colour and form (see pls. 79, 86).

85 *Phlox drummondi*

Seed packet from Carter's Seeds, Raynes Park, London, *c.*1935-40

86 Sweet pea (*Lathyrus odoratus*) 'Blue Lagoon'

Seed packet from Carter's Seeds, Raynes Park, London, *c.*1935-40

Each image focuses on the flower head and is printed in bold flat colour. The packet of 'Blue Lagoon' notes that the plant received an Award of Merit in 1936, as being the finest shade of blue pea then available. The main image of the phlox is supplemented by four diagrammatic flower heads showing the other colours included in the mixed selection.

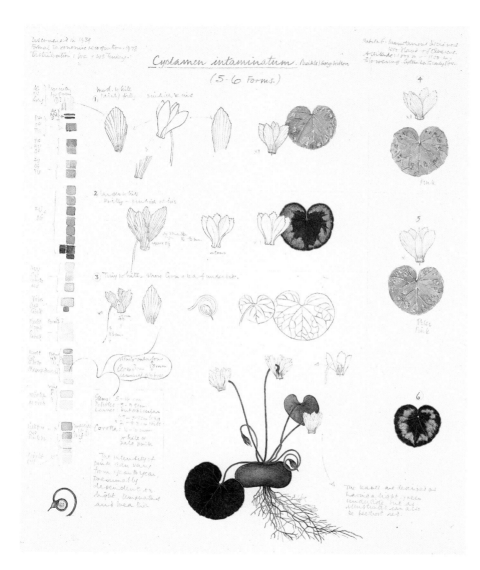

87 Jenny Brasier (b.1936)

Cyclamen intaminatum, **1993**

Pencil and watercolour

This study is related to a plate showing
different forms of species cyclamen painted
for the *Cyclamen Society Journal*. Having
produced several working drawings towards
the finished plate, the artist decided to record
this process by transferring the original work
relating to the different colour forms of one
species – the *Cyclamen intaminatum* – to a
single sheet.

Floras and Field Guides

One of the most common features of botanical illustration, and one rarely questioned or commented upon, is the isolation of the specimen on a blank page. Where and how it grows may be described in the accompanying text, or perhaps indicated in the drawing with a symbolic line or two – as Stella Ross-Craig does with her illustration of the winter aconite by giving a dotted line across the plate to indicate which parts of the plant are above and below the soil. Jacobs[1] notes that when Weiditz illustrated the 'Weissz Seeblum' for Brunfels herbal:

> The plant was taken out of the water, and the roots were cleansed. What therefore we see depicted is a water lily without water – isn't this a bit paradoxical? All relations between the plant and its habitat have been broken and concealed. And yet this is regarded as the first herbal with illustrations 'true to nature'; Weiditz was a pupil of Dürer's and no doubt had learnt from the master the motto about nature: *Wer sie heraus kan reissen, der hat sie* – to tear out nature is to possess her.

Other herbals might include formulaic representations of aquatic plants in water and mention might be made of habitat and locality in those which were moving away from unquestioning dependence on Classical authority (such as Fuchs' *De Historia Stirpium*) but it was only in decorative volumes such as florilegia, and eccentric publications such as *The Temple of Flora*, that the plant portrait had been given a landscape setting. Not until the field guides of the nineteenth and twentieth centuries did illustrations of plants restore them to their native habitat as part of the process of description and identification.

The rise of the amateur naturalist in the nineteenth century saw a concomitant rise in demand for illustrated manuals as an aid to identification when 'botanizing'. Charles Kingsley, writing in 1855, reported that 'books of Natural History are finding their way more and more into drawing rooms and school rooms, and exciting greater thirst for a knowledge which, even twenty years ago, was considered superfluous for all but the professional student'.[2] Natural history 'Field Clubs', and botanical societies were commonplace. Many were of course concerned primarily with cultivated plants and new introductions, but there was also a new interest in the native flora – many important European floras (books devoted to a complete description of the plant life of one specific country or region) appeared or began

1 Jacobs, M. 'Revolutions in plant description', *Liber Gratulatorius in Honorem H. C. De Wit*, ed. J. Arends., G. Boelema, C. de Groot and A. Leeuwenberg, Wageningen, 1980, pp.160-3.

2 Kingsley, *Glaucus, or the wonders of the shore*, 1855, p.7.

88 Simon Verelst (1644-1721)

Tulips

Watercolour

Verelst was a noted painter of flower pieces, decorative amalgamations of flowers that would never bloom simultaneously in nature. To create such a composition the artist was compelled to work piecemeal through the seasons, adding to the piece as flowers became available. Alternatively, and more practically, he would work from sketches such as this which gives two views of a tulip as it might be incorporated into a bouquet. It seems that certain artists would work from a 'library' of sketches, because it is not uncommon to find the same flower repeated in different compositions. These studies, though botanically accurate, show only the flower, and are clearly decorative in intent.

89 William Kilburn (1745-1818)

Primrose (*Primula vulgaris*)

Plate No.107 from William Curtis, *Flora Londinensis*, 1777-98

Engraving, coloured by hand

This plate is not signed, but Kilburn's original watercolour is in the Botany Library of the Natural History Museum. Most of the plates in the first volume were his work. One of the subscribers, Sir Thomas Frankland, wrote often to Curtis with comments and criticisms, but with praise for Kilburn's draughtsmanship. On 21 December 1778 he wrote, 'I heartily wish Kilburn would execute some of the orchis's for you next Spring. There is an uncommon taste in his drawing & I often regret that he has deserted you.'* We see here, as in many plates of the *Flora Londinensis*, a return to the practice of showing the whole plant, complete with roots.

* Curtis Museum, Alton, Hampshire. Curtis letters, item 71-72.

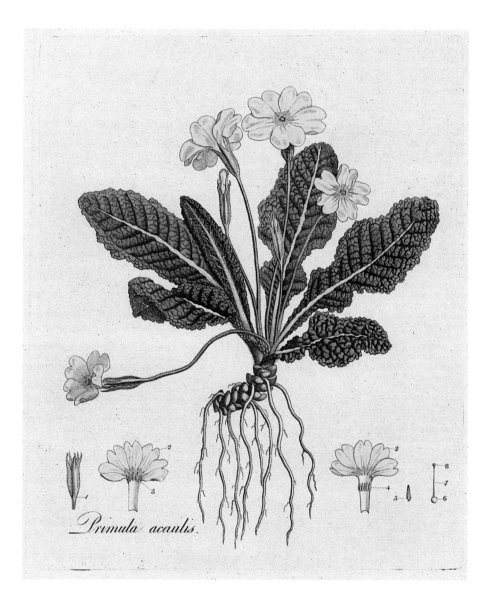

Primula acaulis.

publication in the second half of the eighteenth century or the early nine-teenth: the *Flora Danica* (1762-1871), Sibthorp and Smith's *Flora Graeca* (1806-40), Sowerby's *English Botany* (1790-1814).

In Japan too, botanists began to examine their own flora after centuries of dependence on Chinese plant lore. To some extent this change was stimu-lated by the work of Westerners. Kawahara Keiga (1786-1862) illustrated Japanese plants for Philipp Franz von Siebold, and his drawings are repro-duced, unattributed, in Siebold and Zuccarini's *Flora Japonica* (1835-70). Keiga also began his own flora of Japan, illustrating herbs and trees, in 1836; an enlarged edition, *Somoku Kajitsu Shashin Zafu* [A collection of pictures of plants and fruits] appeared in 1842. Trained by Siebold, Keiga worked in a Western style, using the European graphic syntax. Siebold may also have instructed Yokusai Iinuma who produced a pioneer flora of Japan published as *Shintei Somoku-dzusetsu* [Illustrated book of plants, revised], second edition, 1874, with Latin names given in Roman letters. His plates are an early example of the complete transition to European methods of illustrating, printing and binding.

A more localized flora was William Curtis's *Flora Londinensis* (two vol-umes, 1777-98), describing plants growing within ten miles of London. It had been intended as the first instalment in a comprehensive British flora (and is in fact more inclusive than its title suggests) but it failed through lack of support, and Curtis redirected his energies into the more financially suc-cessful *Botanical Magazine*. The failure of the *Flora Londinensis* is surpris-ing since its illustrations achieved a rarely surpassed level of beauty and accuracy. The engraved plates show the plants life-size, and in the coloured editions (which, at five shillings, cost twice as much as plain) the hand-colouring is unusually delicate and skilful. Curtis employed around 30 people on the colouring, all of whom were closely supervised. The publica-tion also employed some fine botanical artists, notably William Kilburn (1745-1818), later a designer and manufacturer of calicoes which employed exquisite naturalistic floral motifs; and also James Sowerby and Sydenham Edwards.

Curtis took considerable pains with his illustrations for the *Flora Londinensis*. He declares in the introduction his intention 'to have them drawn from the living specimens most expressive of the general habit or appearance of the plant as it grows wild'. This concern for the 'representa-tive' example is shared by the authors of other flora; Stella Ross-Craig was

90 Iinumi Yokusai

Marvel of Peru (*Mirabilis jalapa*)

**From *Somoku-dzusetsu*
[A revised encyclopaedia of botany],
second edition, 1874**

Woodcut

This pioneer flora is detailed and accurate,
with enlarged dissections in the Western
manner, and Latin names given in Roman
letters. It is alleged that Iinumi was instructed
by P. F. Siebold, a German scientist who
co-authored a flora of Japan. He has
nevertheless preserved certain Japanese
graphic characteristics, such as rendering the
leaves white on black for the upper surface,
black on white for the lower.

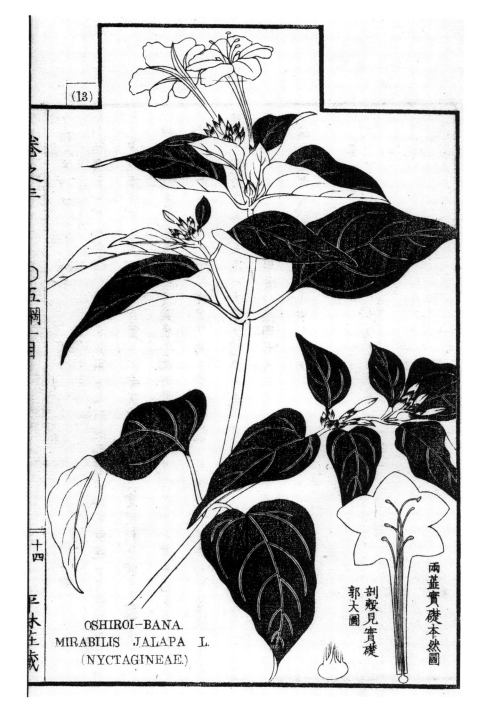

OSHIROI-BANA.
MIRABILIS JALAPA L.
(NYCTAGINEAE.)

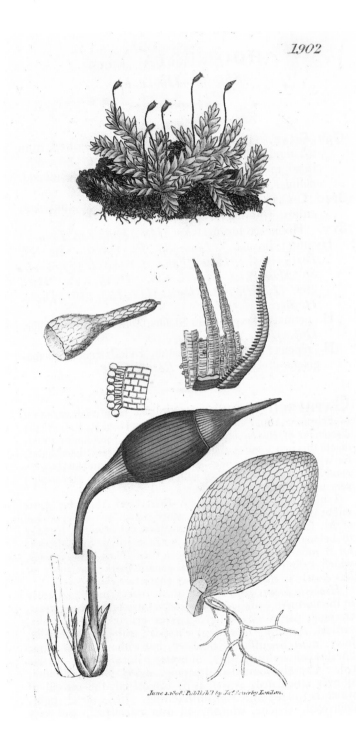

91 James Sowerby

Hookeria lucens

No.1902 from *English Botany*, Volume XVI, 1803

Engraving, coloured by hand

The comprehensiveness of the *English Botany* is apparent in its thorough account of the Cryptogamia, or non-flowering plants, a group which includes ferns, mosses and fungi. The plates illustrating mosses are heavily reliant on magnified details which dwarf the image of the whole plant given life-size.

92 Herb Robert
(*Geranium robertianum*)

Plate 142 from William Curtis, *Flora Londinensis*, London, 1777-98

Engraving, coloured by hand

In almost every case the plants in the *Flora Londinensis* were shown life-size on folio pages large enough not to restrict their subjects. Like the early herbals, and especially the manuscript herbals, Curtis's artists show the whole plant, complete with roots, as an intact entity, and not fragmented as was the predominant mode in scientific illustration of the period. Thus herb Robert is shown in its naturally-spreading habit of growth, but laid out clearly, with very little overlapping of the parts. The image thus combines aspects of the naturalistic and schematic strands in botanical illustration. An unsurpassed delicacy in the hand-colouring, and a keyed floral dissection complete the portrait.

Geranium robertianum

to make similar efforts to identify the typical and characteristic specimen, and it is obviously of considerable importance for a work that aims to be an authoritative source for recognition and identification. Curtis also took some pains with the aesthetics, aiming 'to place each plant, as much as is consistent, in the most pleasing point of view.' Nor is scientific accuracy neglected for he will be 'very particular in the delineation and description of the several parts of the flower and fruit, more especially where they characterise the plant...'

Sowerby's *English Botany* ('or coloured figures of British plants') was a serial publication issued in 267 numbers. The hand-coloured plates were mostly drawn and engraved by Sowerby himself, with botanical descriptions by James Edward Smith. Suffolk botanist Thomas Jenkinson Woodward (1745-1820) wrote to Smith[3] 'I have got the first No. of English botany, and am charmed with it... The paper, print and execution of the plates, is so much superior to Curtis...' Smith himself was somewhat piqued by the general attribution of the success of the *English Botany* to the illustrations. He wondered[4] 'whether the great facility with which a trivial and superficial knowledge of plants is now gained, by turning over books of coloured figures, may not be injurious to true science'. It is not surprising he was resentful, as the overlooked author, but his remarks imply that illustrations are no more than entertainment, of no value unless considered alongside a close reading of the text. But for the amateur, primarily concerned to recognize and to name those plants he encounters in the wild or chooses to grow, pictures are often both his starting point and his end.

The *English Botany* has continued to be an authoritative work of reference to the present day; the plates in Syme's edition (1863-86) are still cited by modern authors including Clapham, Tutin and Warburg in the first, unillustrated, edition of their *Flora of the British Isles* (1952); they are also copied, uncredited, in the present author's copy of the 1962 edition of T. H. Scott and W. J. Stokoe's *Wild Flowers of the Wayside and Woodland*. Surprisingly perhaps, no complete British flora giving both detailed analyses and a life-size portrait of each species appeared until Stella Ross-Craig began work in 1948 on her magnificent series *Drawings of British Plants*. Issued in 32 parts between 1948 and 1973, this work contains 1286 full-page black-and-white plates. They are now the standard illustrations of the British flora, and are particularly valuable for their depiction of seed characters.

3 25 December 1790: *The correspondence and miscellaneous papers of Sir James Edward Smith*, in the Library of the Linnaean Society of London, 18, fol.93.

4 *English flora*, 1, p.xix (1824).

A flora is an invaluable tool in the process of creating a taxonomy, and in identification, but a complete flora is not readily portable on field trips. The field guide appeared to fulfil a particular need: a pocket-sized illustrated guide to the plants of a particular country or region which might be taken into the field and used as reference for immediate recognition. This also had the valuable effect of reducing the need to pick wild flowers or dig them up entire to be identified at home. Curtis himself lamented that the bee orchid was becoming scarce around London because the 'curiosity of Florists... often prompt them to exceed the bounds of moderation, rooting up all they find, without leaving a single specimen to cheer the heart of the Student in his botanical excursions'.

The field guide demands a different kind of work from the illustrator; being small the illustrations must be simple, giving all the information needed for an identification, but with no superfluous details of botanical character. Generally speaking, field-guide illustrations focus on that most distinctive feature, the flower, with fruit, leaves and occasionally, roots, included as supporting evidence to distinguish between similar species. Since they are mostly cheap publications for a general market, the quality of printing is rarely adequate to reproduce a highly-detailed original.

To save space many plants may be included in one plate, as in Keble Martin's *Concise British Flora* (1965). A further economy of space is achieved by showing several plants simultaneously in a single habitat, hence compositions like the fold-out plates by Paxton Chadwick for Cassell's *Pantoscope* natural history series. These also emphasize the purpose of the field guide which, as its name suggests, is as much concerned with habitat and location as with individual plants. Other illustrations use the habitat as the setting for a portrait of a single species, or as a vignette contributing to a complete 'picture' of a plant, as for example in Ann Farrer's plates to *Collins Guide to the Grasses, Sedges, Rushes and Ferns of Britain and Northern Europe* (1984).

The field guide (along with horticultural literature) has seen the most consistent and effective application of photography to botanical illustration. The use of photography in this context has been pioneered by Roger Phillips, beginning with *Wild Flowers of Britain* in 1977. The book is described as an 'encyclopaedia' that can also be used 'in the field'; whilst not pocket-sized it probably lives in many family cars ready for excursions in the countryside. Phillips produced the book as a consequence of his own

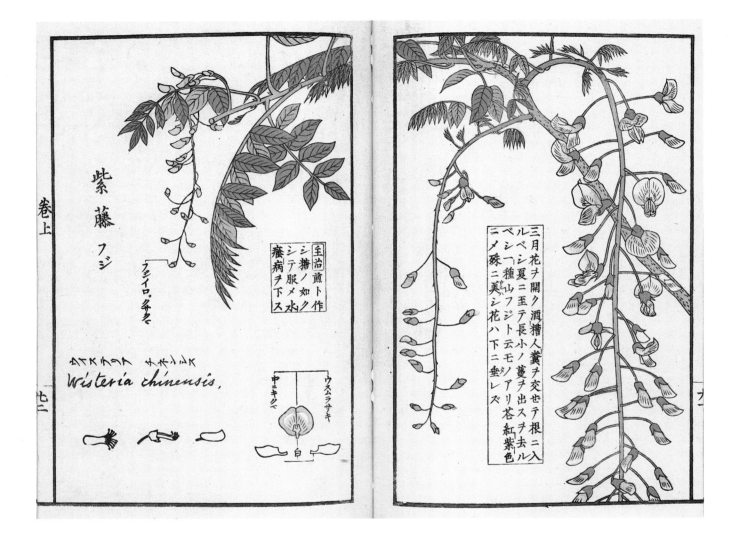

93 Kawahara Keiga

Wisteria sinensis

From *Somoku Kajitsu Shashin Zafu*, **Volume I, 1842**

Woodcut, coloured by hand

Woodcut was the traditional and preferred medium for book illustration in Japan, and despite its proven limitations, it continued to be used for botanical works until the late nineteenth century. Keiga's plates for this flora show that the Japanese were at this period assimilating a Western graphic syntax for the representation of botanical subjects.

94 Roger Phillips

Page 17 (25 March) from *Wild Flowers of Britain*, 1977

Photograph

Phillips sets his plants out in groups on a white ground in an arrangement that recalls the composite pages found in early herbals and plant books such as Parkinson's *Paradisus*.

Since the book is intended as an accessible field guide for the general reader, the plants are grouped not by family but by season of flowering. Thus we have the lesser celandine and stinking hellebore (both Ranunculaceae) with sweet violet (Violaceae) and Primrose (Primulaceae). The violets are shown complete with the roots, the other plants as cut stems or rootless clumps. It is instructive to compare the compact form of Phillips' primrose with that in the *Flora Londinensis* where the artist has 'arranged' and manipulated his specimen to separate and isolate its parts whilst still representing it as an intact whole (pl. 89). The most a photographer can do is search out a typical specimen; he cannot make good its defects or its individuality. However the differences emphasize Phillips' purpose which is less botanical than practical.

curiosity about wild plants, and his frustration with the standard flora or field guide which catalogues plants by family and thus assumes a certain degree of botanical knowledge on the part of the user. As a photographer he set out 'to make a book in which the visual is paramount' and to organize his subjects in a way that is accessible to those completely without botanical expertise. Like the early florilegia, the plants are arranged by season of flowering; some appear again in their fruiting phase, where this is distinctive or characteristic. Cut specimens from related habitats are laid out on a white sheet, and where the root or bulb is distinctive the whole plant is shown. This mode of presentation is reminiscent both of florilegia and the early herbals. The rarest plants are photographed *in situ*. Though photography has its limitations – the fact that Phillips is compelled to photograph some plants twice at different times of the year to give a full identification, where an artist could combine both stages in a single plate – it nevertheless gives quite enough detail for an amateur to make a confident and accurate identification in the field.

95 Stella Ross-Craig (b.1906)

Dog violet (*Viola canina*)

Plate 13, Volume IV of *Drawings of British Plants*, 1948-73

Line-block

Stella Ross-Craig has described her method of drawing a plant and composing a plate as follows:

'When making a water-colour painting of a living specimen, I first study the plant from all angles – as a sculptor might study a head when making a portrait – to grasp its *character*. To understand the structure of a flower it is sometimes necessary to use a magnifying glass. The most pleasing aspect having been decided upon, I sketch in the composition… Leaves are adjusted within scientific reason, to make an agreeable arrangement, and flowers that have been damaged in the post or are for any other reason defective, are replaced in the drawing by perfect specimens. When a flowering stem has to be drawn, and room has to be left for the addition later of leaves and fruits, it may be necessary to consult herbarium specimens in order to judge the space required. At this stage consideration has to be given to the spacing of enlarged dissections of parts of the plant, if these are to be included in the plate. It is, incidentally, a great pity that so little attention is paid nowadays to the underground parts of plants; sixteenth-century artists were well aware of their importance.'

[Quoted in Blunt and Stearn, pp.297-8]

She goes on to express a preference for line drawing because none of the detail is lost in the block-making process. Aside from the magnified details included, Ross-Craig's plates have evident similarities with Fuchs' – the emphasis on outline, with minimal shading; showing the whole plant including the root; and as here, showing the plant twice, in its flowering and fruiting phases.

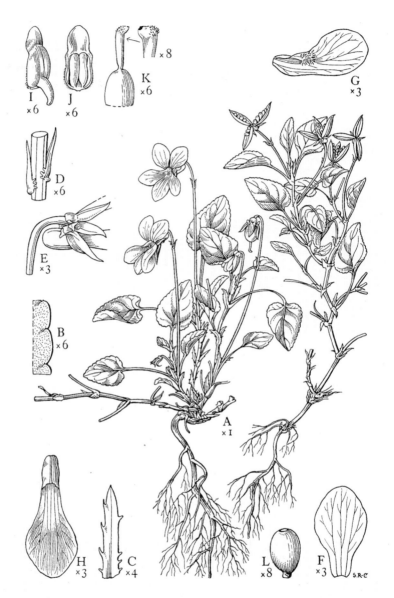

Viola canina L.   Dog-violet
(*V. canina* var. *ericetorum* Reichb.)

A, flowering and fruiting plants; B, margin of leaf, upper surface; C, stipule; D, bracteoles and part of pedicel; E, part of pedicel and basal part of flower; F, upper petal; G, lateral petal; H, lower petal; I, one of the two spurred stamens, back view; J, one of the three upper stamens, front view; K, gynoecium and enlargement of the stigma (the papillae on the latter may be absent); L, seed. Petals clear azure blue.

IV plate 13

137

96 Paxton Chadwick (1903-61)

Flowers of the marsh

Study for Plate 1 for No.11 in the *Pantoscope Series*, Cassell, 1960

Watercolour and crayon

Using a panoramic format, Chadwick shows approximately sixteen species common to marsh landscapes. The habit of growth and the flowering phase are distinctly characterized, but the plate represents a falsification of nature on two levels: it is unlikely that all these plants would be found together on a single site; nor do they all flower at the same time. However, when these conventions are understood, it is an effective and economic mode of picturing plants for a simple field guide.

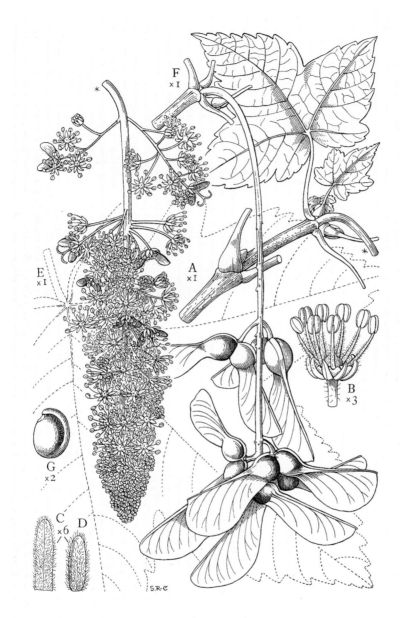

Acer pseudoplatanus L.   Sycamore

A, flowering branch; B, male flower; C, sepal, inner surface; D, petal, inner surface;
E, mature leaf; F, fruiting branch; G, seed.
Sepals and petals pale greenish-yellow. Leaves almost glabrous.

                                                      VI plate 55

97 Stella Ross-Craig (b.1906)

Sycamore (*Acer pseudoplatanus*)

Plate 55, Volume VI of *Drawings of British Plants*, 1948-73

Line-block

This plate exemplifies Ross-Craig's compositional skills. She manages to show the flowering branch (A), fruiting branch (F) and mature leaf (E) life-size within the confines of her 15 x 25.4 cm page, as well as incorporating magnified details of flower and seed, without confusion or distortion. In order to achieve this she uses certain standard conventions from the graphic syntax of botanical illustration: the mature leaf is shown as a broken outline behind the other more fully-realized features, and the flowering branch is cut and presented in two parts, with the point at which they should join indicated by matching asterisks. Ross-Craig works to the very edges of her rectangular plate without allowing the composition to appear constrained or dictated by the format.

The Real Thing:
Herbaria, Nature Prints, Photographs

A good botanical illustration serves as a substitute for its subject, but there are also practices in which the plant itself is preserved as its own record, or is used directly as the physical agent from which an image is made.

In the early sixteenth century, botanists developed a method of preserving plants by drying them pressed between sheets of paper The flattened specimens were then mounted on stiff paper or card. Plants thus preserved retained many of their characteristic features, though there was a loss of volume, texture and colour. Nevertheless they could be stored for permanent reference, used by botanists and by illustrators as source material when the living plant was not available (one of the most valuable skills in a botanical illustrator is the ability to work up an accurate and convincing portrait from a herbarium specimen), and exchanged between botanists. Conrad Gessner, writing from Zurich to Dr. Kentmann in Dresden, says 'if you have any rare plants to be named it is sufficient to send me the dried flower and leaf'.[1] The letters of the Florentine floriculturist Matteo Caccini (1573-1640) to the botanist L'Ecluse and the nurseryman Sweert document the extensive exchange trade in seeds, bulbs and illustrations.

Collections of dried plants became known as 'herbaria' (from *herbarius*, the Latin word for a written herbal). Other terms applied included *hortus hyemalis* (winter garden) and *hortus siccus* (dried garden). Luca Ghini (1490-1556), founder of the Pisa Botanic Garden, compiled what may have been the first herbarium, and his letters record the dispatch of dried specimens to fellow botanists. The first written instructions for pressing and drying plants appear to be those given by Adrian Spieghel in his *Isagoges in Rem Herbarium* (Leiden, 1603). Samuel Pepys describes a visit to John Evelyn (5 November 1665) when 'He showed me his Hortus Hyemalis; leaves laid up in a book of several plants kept dry, which preserve colour, however, and look very finely, better than any Herball'.

Herbaria are still maintained as a vital reference tool for the botanist; important collections exist at the Royal Botanic Gardens, Kew, and in the Natural History Museum, London, where they are supplemented by illustrations. But dried or pressed plants are not confined to the herbarium. A number of writers have used actual specimens to illustrate their books.

1 Quoted in Morton, A. G., *History of Botanical Science*, London, 1981, note 26, 9. 153.

98 Anna Atkins (1799-1871)

Dandelion (*Taraxacum officinale*)

From *Cyanotypes of British and Foreign Flowering Plants and Ferns*, 1854

Atkins was the first to use photographs for illustrated books. Each cyanotype is a hand-made unique print, so none of the plates in Atkins' botanical publications are identical. Despite being an 'impression' of the plant itself, the value of a cyanotype print to the botanist is limited because it can give no indication of volume, texture or colour.

99 After P. J. Redouté

Peruvian apple cactus
(*Cereus peruvianus*)

Plate from A. P. de Candolle, *Plantarum Succulentarum Historia*, Paris, 1799[-1805]

Colour stipple engraving, finished by hand

Illustrations of cacti and succulents are of particular value to the botanist because the difficulties of making herbarium specimens of such plants means that illustrations are often the only record.

This work is the 'Histoire des Plantes Grasses' to which Redouté referred in his remarks on the effectiveness of his colour printing process (see pl. 43).

Obviously some plants lend themselves to this process more readily than others: seaweeds, ferns, and mosses have all been used in this way. A good example is William Gardiner's *Twenty Lessons on British Mosses* '...illustrated with twenty-five specimens' (1852).

As well as preserving the plant itself botanists and artists have attempted to make prints from it. Such images are generally referred to as nature prints. Perhaps the earliest example of the process is the impression of *Salvia officinalis* by that great innovator Leonardo da Vinci, which was made by coating a leaf with lamp-black mixed with oil and then pressing it onto the paper as one would a printing-block[2]. Various botanists experimented with the process; the first to use it on a commercial scale was Johann Hieronymous Kniphof (1704-63) who published *Botanica in Originali* (1747; Linnaean name edition, 1757-64). The limitations of such a process for published works is obvious – the number of impressions that could be made from a single leaf, though variable, were unlikely to run to more than 30 for simple sturdy plants, and many fewer for those more complex and fragile. Johann Beckmann compares such impressions with the 'most perfect representations of plants in engravings elegantly coloured' by Ehret, Miller and others. In his view nature prints 'exceed the former in this respect, that they express better some of the internal prominent parts, fibres, veins, &c' but 'on the other hand, they exhibit only the contour of dead and bruised plants, whereas the former present the living image'.[3] In other words the simulacrum carries more conviction, is more 'real', than an impression of the thing itself, in terms of its value as a complete picture.

The technological innovations of the nineteenth century presented a means of transcending this particular limitation when it became possible to make a mould from the plant and to use this as the printing block. There is some debate as to who originated the process but in the early 1850s the Imperial Printing Office in Vienna devised a method which involved sandwiching a plant between a smooth sheet of copper and another of lead, and putting it into a press. The impression of the plant formed in the soft lead was used to make an electrotype for printing. Henry Bradbury patented a similar process in England in 1853, using steel and copper respectively. The botanist John Lindley described its value to botanical publication: 'an exact copy in copper of the part to be represented being employed by the printer instead of so fragile an object as the plant itself, we obtain the means of multiplying copies to the same extent as in copperplate engraving; and hence the method becomes suitable for purposes of publication'.[4]

2 In the Biblioteca Ambrosiana, Milan.

3 Beckmann, J. *A History of Inventions...translated from the German by William Johnston*, 2nd ed., London, 1814, vol.4, pp.621-9.

4 Quoted in Brent Elliott, 'Lasting Impressions', *The Garden*, October 1993, p.477.

In Bradbury's process the mould was inked with colours as appropriate before printing, and then mould and paper were passed through a roller press. The result was an embossed impression, with all the colours printed in one pressing. Bradbury produced a selection of specimen pages in 1854; his first book was *The Nature-Printed British Ferns* (1855-6) with text by Thomas Moore, then Curator of the Chelsea Physic Garden, and edited by John Lindley. Lindley praised the process as an improvement on the old method because it represented 'not only general form with absolute accuracy, but also surface hairs, veins, and all those minutiae of superficial structure by which plants are known'. This was nevertheless limiting, as Lindley acknowledged, but as he concedes in his preface 'its accuracy is perfect as far as it goes; and in the case of British Ferns it goes far enough for all practical purposes'.

Of course the most profound limitation of nature-printing was that it could only produce accurate impressions of certain kinds of plants. Reviewing Johnstone and Croall's *The Octavo Nature-Printed British Seaweeds* (1855-6), the *Gardener's Chronicle* in 1859 admired the process and remarked that 'when flat thin bodies are to be represented, such as the leaves of plants, the human hand is wholly superseded'. The most suitable candidates for nature-printing were precisely those that were thin and flat; ferns were especially good, but successful results were also achieved by Bradbury with the lime tree and the primrose. It was plainly inadequate for fleshy, delicate plants such as the pasque flower and the crocus, as demonstrated in one of Bradbury's specimen plates (entitled 'Phytoglyphy or The Art of Printing From Nature') where three plants are modelled in their entirety. Generally leaves and stems are more clearly defined than petals, but the colouring also tended to obscure the finer details.

Bradbury died in 1860, and although a variation of the nature-printing process, 'phytoxigraphique' was used to illustrate a French flora in the 1860s and 1870s, the process was effectively superseded, not least by photography, itself a mode of illustration that might also be defined as a reproduction of the thing itself. One of the first people to apply photographic processes to a serious botanical purpose, albeit as an amateur, was Anna Atkins (1797-1871). Atkins's father, John Children, was Vice President of the Botanical Society of London; Atkins was herself elected a member in 1839; on 1 March, 1839, John Thomas Cooper, Jr. showed members of the Society 'numerous figures of Mosses and Ferns produced by the Photogenic process of Mr. Talbot'[5]. Fox Talbot's method of 'photogenic drawing' was

5 *Annals of Natural History*, Vol.4, no.23, November 1839, p.212

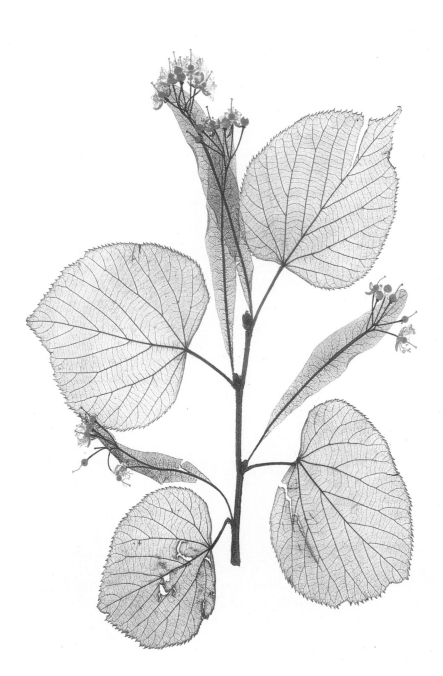

100 Bradbury and Evans

Common lime or linden (*Tilia* x *vulgaris*), 1854

Nature print

Henry Bradbury's first efforts in nature printing were issued in 1854 as a volume of 21 plates entitled *A few leaves represented by 'nature printing' showing the application of the art for the reproduction of botanical and other natural objects with a delicacy of detail and truthfulness unobtainable by any other known method of printing…* The 'book' had no text and seems to have been intended as an advertisement for the potential applications of nature printing. It was quickly followed by Moore's *Ferns of Great Britain and Ireland* (1855), and Johnstone and Croall's *Nature-printed British Sea-weeds* (1859) (see pl. 101).

Although modelled on the actual plant, nature prints lack the illusion of three dimensions conveyed by a conventional illustration, and in fact resemble flattened herbarium specimens.

Plate LXXVI

CHYLOCLADIA clavellosa, AG.

Nature Printed by Henry Bradbury.

101 *Chylocladia clavellosa*

Plate LXXVI from Johnstone and Croall
***The Nature-Printed British Sea-weeds*,**
Volume II, 1859

Nature print

The nature print of the plant itself is
supplemented with outline drawings (on the
same plate) of magnifications of its structural
characteristics including tetraspores and
sections of the stem.

adapted by astronomer and scientist Sir John Herschel who discovered in 1842 that paper impregnated with iron salts was light-sensitive, and on exposure to light, prussian blue pigment was formed in the paper. It was this, the cyanotype or 'blueprint' process, that was adopted by Atkins for her botanical recording.

Her father had written to botanist William Hooker in 1835, saying that 'my daughter, Mrs. Atkins, has a great fondness for Botany, and is making an Herbarium'.[6] It was but a small step from preserving the specimens themselves, to preserving their exact likeness. In 1843 she began using the cyanotype process to record material from her herbarium, issuing the first instalment of *British Algae* in October 1854. She explains in the introduction that 'the difficulty of making accurate drawings of objects as minute as many of the Algae and Conferva, has induced me to avail myself of Sir John Herschel's beautiful process of Cyanotype, to obtain impressions of the plants themselves'.[7] She was to spend ten years on the project, producing more than a dozen copies of a publication that contained nearly 400 original plates. However, cyanotype had the advantage of being one of the simplest photographic processes: to produce a print the plant specimen was laid on a sheet of sensitized paper and exposed to sunlight for a few minutes. This produced a faint image which was enhanced to a rich blue and made permanent by washing the paper in water. Though these cyanotypes were undoubtedly a valuable means of recording plants such as algae, with their minute filaments, there were obvious limitations for their botanical application of the process. First of all, the plant was represented in silhouette, without even the surface detail that a nature print included, and second, the illustration was always of a unique individual specimen with all the potential drawbacks that entailed for botanical theory and identification.

Both the cyanotype and its variant the photogenic drawing were dead ends for botanical illustration, but photography itself went on to become the dominant medium in virtually every field of illustration. Photography allied with other technological tools – microscopes, scanners, computers – has allowed science to explore aspects of the physical nature of plants hitherto invisible. Photomicrography for instance has proved valuable in giving us magnified details of a plant's molecular structure. Nevertheless photography is even now unable to fulfil all the needs of the botanist; botanical draughtsmanship survives as a distinct discipline because it has the flexibility to focus, analyze and dissect its subject, and to combine disparate parts in a clearly intelligible design.

6 Letter of 3 October 1835, Royal Botanic Gardens, Kew.

7 Atkins was herself a skilled scientific draughtsman who had contributed the drawings of shells to her father's translation of Lamarck's *Genera of Shells* (1823).

Bibliography

Arber, Agnes, *Herbals: Their Origin and Evolution*, Cambridge, 1938.

Archer, Mildred, *Natural History Drawings in the India Office Library*, London, 1972.

Barker, Nicholas, *The Hortus Eystettensis: The Bishop's Garden and Besler's Magnificent Book*, London, 1994.

Bartlett, H. H. and Shohara, H., *Japanese Botany During the Period of Wood-block Printing*, Los Angeles, 1961.

Blunt, Wilfred and Raphael, Sandra, *The Illustrated Herbal*, London, 1975.

Blunt, Wilfred and Stearn, William T., *The Art of Botanical Illustration*, revised ed., 1994.

Calman, Gerta, *Ehret, Flower Painter Extraordinary*, Oxford, 1977.

Desmond, Ray, *Wonders of Creation, Natural History Drawings in the British Library*, London, 1986.

Desmond, Ray, *A Celebration of Flowers: Two Hundred Years of Curtis's Botanical Magazine*, Kew and Twickenham, 1987.

Ehret, G. D., 'A Memoir of Georg Dionysius Ehret...written by himself', *Proceedings of the Linnaean Society*, London, 1894-5, pp.41-58.

Henrey, Blanche, *British Botanical and Horticultural Literature before 1800*, London, 1975.

Hulton, Paul, *The Work of Jacques Le Moyne de Morgues*, 2 vols., London, 1977.

Hulton, Paul and Smith, Lawrence, *Flowers in Art from East and West*, London, 1979.

Hunt, J. Dixon and Jong, E. de, Special double issue of 'The Anglo-Dutch Garden in the Age of William and Mary', *Journal of Garden History*, Vol. 8, nos. 2 and 3, 1988.

Kaden, Vera, *The Illustration of Plants and Gardens 1500-1850*, London (V&A Museum), 1983.

Mabey, Richard, *The Flowering of Kew: 350 Years of Flower Painting from the Royal Botanic Gardens*, London, 1988.

Prest, John, *The Garden of Eden: the Botanic Garden and the Re-Creation of Paradise*, New Haven and London, 1981.

Rix, Martyn, *The Art of Botanical Illustration*, London, 1981.

Royal Horticultural Society, *Proceedings of the First Symposium of Botanical Art*, 1993.

Scrase, David, *Flowers of Three Centuries: One Hundred Drawings & Watercolours from the Broughton Collection*, Washington D.C., 1983.

Sotheby's, *The Glory of the Garden* (catalogue of a loan exhibition in association with the Royal Horticultural Society), London, 1987.

Stearn, William T., *Flower Artists of Kew*, London, 1990.

Acknowledgements

I would like to thank all those who have contributed to the production of this book. I have benefitted from the ideas, advice and support of a great many people, not least my colleagues in the collection of Prints, Drawings and Paintings at the Victoria and Albert Museum. I am especially grateful to Sharon Fermor, Chris Titterington and Charles Newton, all of whom read the first draft and offered valuable comments. The clarity and accuracy of the text has been much improved by the careful and creative editing of Georgina Harding; any errors of fact or infelicities of expression which remain are my own. I must also thank the staff of the National Art Library, Rebecca Coombes in particular, for their patience and diligence in facilitating my research; Greg Irvine of the Far Eastern Collection for identifying much relevant Japanese and Chinese material; Amanda Robertson, Moira Thunder and Tim Travis for organizing all the photography; and Philip Spruyt de Bay, Ian Thomas and Christine Smith for the excellent photographs.

Finally I must acknowledge my huge debt to all those whose researches I have drawn upon in the writing of this book. Most are credited in the bibliography or the footnotes, but I must add a special mention of Roger Phillips and Jenny Brasier, both for their generous gifts to the Museum, and for sharing with me fascinating accounts of their working practices. I must also thank Ute Krebs for her invaluable research on Walther, and her translations of the text of the Nassau florilegium.

All the works illustrated are in the collections of the Victoria and Albert Museum and are reproduced by courtesy of the Board of Trustees.

Index

VNA HIERVNDA

Omnia VERE vigent,
et totus fervet V